**Progress in Nonlinear Differential Equations and Their Applications**
Volume 18

J. J. Duistermaat

# The Heat Kernel Lefschetz Fixed Point Formula for the Spin-c Dirac Operator

Birkhäuser
Boston · Basel · Berlin

J. J. Duistermaat
Mathematisch Instituut
Universiteit Utrecht
3508 TA Utrecht
The Netherlands

**Library of Congress Cataloging-in-Publication Data**

Duistermaat, J. J. (Johannes Jisse), 1942-
The heat kernel Lefschetz fixed point formula for the spin-c dirac operator / J. J. Duistermaat
p. cm. -- (Progress in nonlinear differential equations and their applications ; v. 18)
Includes bibliographical references and index.

1. Almost complex manifolds. 2. Operator theory. 3. Dirac equation. 4. Differential topology. 5. Mathematical physics.
I. Title. II. Series.
QC20.7.M24D85 1995 95-25828
515'.7242--dc20 CIP

Printed on acid-free paper

*Birkhäuser*



ISBN-13: 978-1-4612-5346-4 e-ISBN-13: 978-1-4612-5344-0
DOI: 10.1007/978-1-4612-5344-0

Typeset from author's disk by TeXniques, Boston, MA
Printed and bound by Quinn-Woodbine, Woodbine, NJ

9 8 7 6 5 4 3 2 1

# Contents

# Preface

When visiting M.I.T. for two weeks in October 1994, Victor Guillemin made me enthusiastic about a problem in symplectic geometry which involved the use of the so-called spin-c Dirac operator. Back in Berkeley, where I had spent a sabbatical semester[1], I tried to understand the basic facts about this operator: its definition, the main theorems about it, and their proofs. This book is an outgrowth of the notes in which I worked this out. For me this was a great learning experience because of the many beautiful mathematical structures which are involved.

I thank the Editorial Board of Birkhäuser, especially Haim Brézis, for suggesting the publication of these notes as a book. I am also very grateful for the suggestions by the referees, which have led to substantial improvements in the presentation. Finally I would like to express special thanks to Ann Kostant for her help and her prodding me, in her charming way, into the right direction.

J.J. Duistermaat

Utrecht, October 16, 1995.

[1]Partially supported by AFOSR Contract AFO F 49629-92

# Chapter 1

# Introduction

## 1.1 The Holomorphic Lefschetz Fixed Point Formula

Let $M$ be an almost complex manifold of real dimension $2n$, provided with a Hermitian structure. Furthermore, let $L$ be a complex vector bundle over $M$, provided with a Hermitian connection. We also assume that $K^*$, the dual bundle of the so-called canonical line bundle $K$ of $M$, is provided with a Hermitian connection. We write $E$ for the direct sum over $q$ of the bundles of $(0, q)$-forms; in it we have the subbundle $E^+$ and $E^-$, where the sum is over the even $q$ and odd $q$, respectively. Write $\Gamma$ and $\Gamma^{\pm}$ for the space of smooth sections of $E \otimes L$ and $E^{\pm} \otimes L$, respectively. From these data, one can construct a first order partial differential operator $D$, the spin-c Dirac operator mentioned in the title of this book, which acts on $\Gamma$. The restriction $D^+$ of $D$ to $\Gamma^+$ maps into $\Gamma^-$, and the restriction $D^-$ of $D$ to $\Gamma^-$ maps into $\Gamma^+$. If $M$ is compact, then the fact that $D$ is elliptic implies that the kernel $N^{\pm}$ of $D^{\pm}$ is finite-dimensional, and the difference $\dim N^+ - \dim N^-$ is equal to the *index* of $D^+$.

The Atiyah-Singer index theorem applied to this case [7, Theorem (4.3)] expresses this index as the integral over $M$ of a characteristic class in the De Rham cohomology of $M$, equal to the product of the Todd class of the tangent bundle of $M$, viewed as a complex vector bundle over $M$, and the Chern character of $L$. These characteristic classes are given by polynomial expressions in the curvature forms of the given bundles. If $M$ is a complex analytic manifold, then the index of $D^+$ is equal to the Riemann-Roch number of $M$, and the integral formula generalizes the one which Hirzebruch [39] obtained for complex projective algebraic varieties.

If $\gamma$ is a bundle automorphism of $L$ which leaves all the given structures invariant, then it induces an operator in $\Gamma$ which commutes with $D$, and one can form the *virtual character*

$$\chi(\gamma) = \mathrm{trace}_{\mathbf{C}}\, \gamma\big|_{N^+} - \mathrm{trace}_{\mathbf{C}}\, \gamma\big|_{N^-}. \tag{1.1}$$

Under the assumption that the fixed point set $M^\gamma$ of $\gamma$ in $M$ locally is a smooth almost complex submanifold and that the action of $\gamma$ in the normal bundle is nondegenerate, the equivariant index theorem of Atiyah-Segal and Atiyah-Singer expresses the virtual character as the sum over the connected components $F$ of $M^\gamma$, of similar characteristic classes of the $F$'s. In the complex analytic case, this is called the *holomorphic Lefschetz fixed point formula*, cf. Atiyah and Singer [7, Theorem (4.6)]. In the case of isolated fixed points, it is due to Atiyah and Bott [5, Theorem 4.12].

## 1.2 The Heat Kernel

The operator $Q^+ = D^- \circ D^+$ maps $\Gamma^+$ to $\Gamma^+$, and $Q^- = D^+ \circ D^-$ maps $\Gamma^-$ to $\Gamma^-$. Each of the operators $Q^+$ and $Q^-$ is equal to a Laplace operator, plus a zero order part which involves curvature terms. The corresponding heat diffusion operators $\mathrm{e}^{-tQ^\pm}$ are integral operators with a smooth integral kernel $K^\pm(t,\, x,\, y)$, $t > 0$, $x,\, y \in M$. Along the diagonal $x = y$, and for

$t \downarrow 0$, these kernels have an asymptotic expansion of the form

$$K^{\pm}(t,\, x,\, x) \sim t^{-n} \sum_{k=0}^{\infty} t^k \, K_k^{\pm}(x). \tag{1.2}$$

In this asymptotic expansion, each of the coefficients $K_k(x)^{\pm}$ is given by a universal polynomial expression in a finite part of the Taylor expansion of the geometric data at the point $x$.

It was observed by McKean and Singer [57, p. 61] that

$$\operatorname{index} D^{+} = \int_M \operatorname{trace}_{\mathbf{C}} K_n^{+}(x) - \operatorname{trace}_{\mathbf{C}} K_n^{-}(x) \,\mathrm{d}\, x, \tag{1.3}$$

and they asked the question if not, by some fantastic cancellation, the higher order derivatives in the expression for $K_n^{\pm}(x)$ cancel, to give that the integrand in (1.3) is equal to a characteristic differential form whose cohomology class is equal to the one of the index theorem. This would give a direct analytic proof of the index theorem, with the advantage of having a local interpretation of the integrand. Actually, in [57] the question is asked for the Euler characteristic of $M$, but it obviously can be generalized to arbitrary index problems.

## 1.3 The Results

It turned out that, also in the presence of an automorphism $\gamma$, the fantastic cancellation indeed takes place. See Theorem 11.1 and Theorem 12.1. In the complex analytic case, the result is referred to as a *local holomorphic Lefschetz fixed point formula*. It is the purpose of this book, to explain both all the ingredients in the formula, and how the answer comes about. In it, we will apply the methods of Berline, Getzler and Vergne [9, Ch. 1-6], and show how these work in the case of the spin-c Dirac operator. (For the comparison: our $L$ is their $\mathcal{W}$, the letter $W$ is the classical notation of Hirzebruch [39]. We have chosen the letter $L$, because of the connotation of a "linear system".)

For the index, the result is due to Patodi [64] in the Kähler case, with another proof by Gilkey [28], who in [31] extended the result to almost complex manifolds. In the presence of an automorphism $\gamma$, the local formula had been obtained by Patodi [65] under the assumption that the connected components of the fixed point set $M^\gamma$ of $\gamma$ in $M$ are Kähler manifolds. A proof for general almost complex manifolds has been indicated by Kawasaki [45, pp. 156-158]. One can also obtain the result in this general setting as a consequence of the local Lefschetz formula for the spinor Dirac operator of Berline and Vergne [11], cf. [9, Theorem 6.11]. That is, by using the comparison (6.20) between the bundle $E$ of $(0,\, q)$-forms and the spinor bundle $S$, and observing that it suffices to work locally, where spin structures always exist.

The local formula is particularly suited for the generalization of the Lefschetz formula to compact orbifolds, which we will explain in Chapter 14. I learned this from the proof of Kawasaki [45] for the Riemann-Roch number. For arbitrary elliptic operators on compact orbifolds, the Lefschetz formula has been obtained by Vergne [73]. She used the theory of transversally elliptic operators of Atiyah [2], as Kawasaki [46] did in his proof of the index formula for orbifolds. The use of the local formula avoids the use of the commutative algebra of [2], which may make it more accessible to analysts.

Strictly speaking, this work contains no new results. However, the spin-c Dirac operator is a very important special case among the general Dirac-type operators. As described above, it came originally from the study of complex analytic manifolds. On the other hand, every symplectic manifold (phase space in classical mechanics) also carries an almost complex structure and hence a corresponding spin-c Dirac operator. We will discuss the application of the theory to this case in Chapter 15. As a third application, we mention that recently the Seiberg-Witten theory, an $S^1$ gauge theory which uses the spin-c Dirac operator, has led to striking progress in the differential topology of four-dimensional compact oriented manifolds. Here one works with spin-c Dirac operators which are defined in terms of spin-c structures which do not necessarily come from an almost complex structure. See the Remark in front of Lemma 5.5. For an exposition of Seiberg-Witten theory see for instance Eichhorn and Friedrich [26] or Morgan [61]. The importance of the spin-c Dirac operator makes it worthwhile to work out the beautiful constructions of [9] for this special kind of Dirac operator.

A large part of the exposition has a wider scope than just the spin-c Dirac operator. For instance, Chapter 8 is an exposition of the asymptotic expansion of heat kernels for generalized Laplace operators, following [9, Ch. 2]. Chapters 9 and 10, on the Berline-Vergne theory of heat kernels on principal bundles, are also written for more general operators than only the spin-c Dirac operator. The point of this theory is, that it gives an explanation for the similarity between the factor $\det \frac{1-e^{-R}}{R}$, which appears in the index formula, and the Jacobian of the exponential mapping from a Lie algebra to the Lie group. (Here $R$ denotes curvature.) Lemma 9.5 and Lemma 9.6 form the starting point of this explanation. Although in general we tried to keep our notations close to our main reference [9], we apologize that at some points we ended up with a different choice.

Finally, in Chapter 13 the formulas of Theorem 11.1 and Theorem 12.1 are translated into the language of characteristic classes, in which the formulas of Hirzebruch and Atiyah-Singer originally were phrased. We use the occasion to explain, in Chapter 16, the Weil homomorphism in its natural setting of equivariant differential forms in the presence of an action of a Lie group, and under the assumption that the action admits a connection form.

I am very grateful to Victor Guillemin for arousing my interest in the subject, in connection with the question how the Riemann-Roch number of a reduced phase space for a torus action is related to multiplicities of intermediate phase spaces. And I apologize for spending so much time on writing up this text, instead of "adorning the dendrites". Finally I would like to thank the Department of Mathematics of UC Berkeley, for providing me with an ideal environment to work on this.

# Chapter 2

# The Dolbeault-Dirac Operator

In this chapter we set the stage, by introducing complex and almost structures, the Dolbeault complex and Hermitian structures. The holomorphic Lefschetz number, defined as the alternating sum of the trace of the automorphism acting on the cohomology of the sheaf of holomorphic sections, will be expressed in terms of a selfadjoint operator, which is built out of the Dolbeault operator and its adjoint; the Dolbeault-Dirac operator in the title of this chapter. This material is very well-known but, also in order to fix the notations, we have taken our time for the description of these structures. Just for convenience, we will assume that all objects are smooth (infinitely differentiable).

## 2.1 The Dolbeault Complex

Let $M$ be a manifold of even dimension $2n$, provided with an almost complex structure $J$. That is, for each $x \in M$, $J_x$ is a real linear transformation in $\mathrm{T}_x\, M$ such that ${J_x}^2 = -1$. A real linear mapping $A$ from $\mathrm{T}_x\, M$ to a complex vector space $V$ is called complex linear and complex antilinear with respect to the complex structure $J_x$ in $\mathrm{T}_x\, M$, if

$$A\,(J_x(v)) = i\,A(v),\;\; v \in \mathrm{T}_x\, M \tag{2.1}$$

and

$$A\,(J_x(v)) = -i\,A(v),\;\; v \in \mathrm{T}_x\, M, \tag{2.2}$$

respectively.

The space of complex linear and complex antilinear forms ($V = \mathbf{C}$) on $\mathrm{T}_x\, M$ is denoted by $\mathrm{T}_x^*\, M^{(1,\,0)}$ and $\mathrm{T}_x^*\, M^{(0,\,1)}$, respectively. With this notation, the space of complex linear and antilinear mappings from $\mathrm{T}_x\, M$ to $V$ becomes equal to $\mathrm{T}_x^*\, M^{(1,\,0)} \otimes V$ and $\mathrm{T}_x^*\, M^{(0,\,1)} \otimes V$, respectively. One has the complementary projections

$$\pi^{(1,\,0)} : \xi \mapsto \xi^{(1,\,0)} := \tfrac{1}{2}\,(\xi - i\,\xi \circ J_x)\,, \tag{2.3}$$

and

$$\pi^{(0,\,1)} : \xi \mapsto \xi^{(0,\,1)} := \tfrac{1}{2}\,(\xi + i\,\xi \circ J_x)\,, \tag{2.4}$$

from $\mathrm{T}_x^*\, M \otimes \mathbf{C}$ onto $\mathrm{T}_x^*\, M^{(1,\,0)}$ along $\mathrm{T}_x^*\, M^{(0,\,1)}$, and from $\mathrm{T}_x^*\, M \otimes \mathbf{C}$ onto $\mathrm{T}_x^*\, M^{(0,\,1)}$ along $\mathrm{T}_x^*\, M^{(1,\,0)}$, respectively.

A complex-valued function $f$ on $M$ is called complex-differentiable or complex-analytic, or holomorphic, if, for every $x \in M$, $\mathrm{d}f_x$ is complex linear. If $\bar{\partial} = \pi^{(0,\,1)} \circ \mathrm{d}$ denotes the operator d followed by the projection (2.4), then this condition is equivalent to the differential equation $\bar{\partial} f = 0$. One also writes $\partial = \pi^{(1,\,0)} \circ \mathrm{d}$, so that $\mathrm{d} = \partial + \bar{\partial}$ on functions, and $\partial f = \mathrm{d}f$ if and only if $f$ is holomorphic.

Let $p,\, q,\, r \in \mathbf{Z}_{\geq 0}$, with $p + q = r$. A complex-valued antisymmetric $r$-linear form on $\mathrm{T}_x\, M$ is called of type $(p,\, q)$, if it is equal to a finite sum of forms $\alpha \wedge \beta$, where $\alpha \in \Lambda^p\, \mathrm{T}_x^*\, M^{(1,\,0)}$ and $\beta \in \Lambda^q\, \mathrm{T}_x^*\, M^{(0,\,1)}$. The space of forms of type $(p,\, q)$ is denoted by $\mathrm{T}_x^*\, M^{(p,\,q)}$. The point is that

$$\Lambda^r\, \mathrm{T}_x^*\, M \otimes \mathbf{C} = \bigoplus_{p,\,q,\,p+q=r} \mathrm{T}_x^*\, M^{(p,\,q)}, \tag{2.5}$$

so we have the projection $\pi_{p,q}$ from $\Lambda^r \mathrm{T}_x^* M \otimes \mathbf{C}$ onto $\mathrm{T}_x^* M^{(p,q)}$ along the sum of the other components. An $L$-valued version is obtained by tensoring $\mathrm{T}_x^* M^{(p,q)}$ with $L_x$. A $(p, q)$-form $\omega_x$ on $\mathrm{T}_x M$, which depends smoothly on $x \in M$, is called a $(p, q)$-form on $M$. The space of $(p, q)$-forms on $M$ is denoted by $\Omega^{(p,q)}(M)$.

In particular we will be interested in the case $p = 0$, for which we will use the following abbreviation throughout:

$$E_x^q := \mathrm{T}_x^* M^{(0,q)} = \Lambda^q \mathrm{T}_x^* M^{(0,1)}, \quad E_x^0 = \mathbf{C}. \tag{2.6}$$

Note that $E_x^q = 0$ if $q > n$, and $\dim_{\mathbf{C}} E_x^q = \binom{n}{q}$ if $0 \leq q \leq n$. We will write

$$E_x := \bigoplus_{q=0}^{n} E_x^q, \tag{2.7}$$

$$E_x^+ = E_x^{\text{even}} = \bigoplus_{\text{even } q} E_x^q, \tag{2.8}$$

$$E_x^- = E_x^{\text{odd}} = \bigoplus_{\text{odd } q} E_x^q. \tag{2.9}$$

With the exterior product of forms and the splitting in $E_x^+$ and $E_x^-$, $E_x$ is a supercommutative superalgebra over $\mathbf{C}$. (See [9, Section 1.3] for the definition of such algebras.)

The $E_x$, $x \in M$, form a complex vector bundle $E$ over $M$ with subbundles $E^+ = \bigcup_{x \in M} E_x^+$ and $E^- = \bigcup_{x \in M} E_x^-$. The space of sections of $E$, $E^+$ and $E^-$ is equal to the direct sum of the spaces $\Omega^{(0,q)}$, where $q$ runs over all the even and the odd integers $0 \leq q \leq n$, respectively.

In Chapter 5 we will introduce the spin-c Dirac operator, which will be used in the general case of an almost complex structure. In order to motivate its definition and to understand its relation to complex analysis, we assume in the remainder of this chapter that $M$ is a complex analytic manifold. This means that around every $x \in M$ there is a system of local coordinates in which $J$ is equal to the standard complex structure of $\mathbf{C}^n$. This is equivalent to the condition that, at every $x \in M$, there exist $n$ holomorphic functions $z_j$ in a neighborhood of $x$ in $M$, such that the $\mathrm{d}z_j$ at $x$ are linearly independent over $\mathbf{C}$.

In such coordinates $z_j$, each $(p,\, q)$-form is of the form

$$\omega = \sum_{J,\,K} \omega_{J,\,K}\; \mathrm{d}z_J \wedge \mathrm{d}\bar{z}_K, \tag{2.10}$$

where $J$ and $K$ runs over the set of strictly increasing sequences $J = (j_i)_{i=1}^p$ and $K = (k_i)_{i=1}^q$, respectively, each $\omega_{J,\,K}$ is a complex-valued function, and

$$\mathrm{d}z_J = \mathrm{d}z_{j_1} \wedge \mathrm{d}z_{j_2} \wedge \ldots \wedge \mathrm{d}z_{j_p}, \tag{2.11}$$

$$\mathrm{d}\bar{z}_K = \mathrm{d}\bar{z}_{k_1} \wedge \mathrm{d}\bar{z}_{k_2} \wedge \cdots \wedge \mathrm{d}\bar{z}_{k_q}. \tag{2.12}$$

From this we see that $\mathrm{d}\omega$ is the sum of a $(p+1,\, q)$-form and a $(p,\, q+1)$-form. Or, one again has $\mathrm{d} = \partial + \bar{\partial}$, if one writes

$$\partial = \pi^{(p+1,\, q)} \circ \mathrm{d} \tag{2.13}$$

and

$$\bar{\partial} = \pi^{(p,\, q+1)} \circ \mathrm{d} \tag{2.14}$$

on $(p,\, q)$-forms.

This implies that for each $(0,\, 1)$-form $\omega$, $\pi^{(2,\,0)}\, \mathrm{d}\omega = 0$. For a general almost complex structure $J$, it need no longer be true that $\mathrm{d} = \partial + \bar{\partial}$. For each $x \in M$, one has the antisymmetric bilinear mapping $[J,\, J]_x$ from $\mathrm{T}_x\, M \times \mathrm{T}_x\, M$ to $\mathrm{T}_x\, M$, which is defined by

$$[J,\, J](v,\, w) = [Jv,\, Jw] - J\,[Jv,\, w] - J\,[v,\, Jw] - [v,\, w], \tag{2.15}$$

for any vector fields $v$ and $w$ in $M$. Using the formula

$$(\mathrm{d}\omega)(v,\, w) = v\,\omega(w) - w\,\omega(v) - \omega\left([v,\, w]\right), \tag{2.16}$$

one gets for each (0, 1)-form $\omega$ that

$$\mathrm{d}\omega(v - i\,Jv,\, w - i\,Jw) = \omega\left([J,\, J](v,\, w)\right).$$

So the condition that $\pi^{(2,0)}\,\mathrm{d}\omega = 0$ for every $(0,\, 1)$-form $\omega$ is equivalent to the condition that $[J,\, J] = 0$. The *theorem of Newlander and Nirenberg* now says that an almost complex manifold $(M,\, J)$ is complex analytic if and only if $[J,\, J] = 0$. This theorem is already valid if the first order derivatives of $J$ are Hölder-continuous. Cf. Newlander and Nirenberg [62], Hörmander [41], Malgrange [55].

We continue the discussion of complex analytic manifolds. Identifying the types in $0 = \mathrm{d}^2 = \partial^2 + \partial\,\bar\partial + \bar\partial\,\partial + \bar\partial^2$ on $\Omega^{(p,\,q)}$-forms, one sees that $\partial^2 = 0$, $\partial\,\bar\partial + \bar\partial\,\partial = 0$, and $\bar\partial^2 = 0$. In particular, the operator $\bar\partial$ defines a complex

$$0 \longrightarrow \Omega^{(p,\,0)} \xrightarrow{\bar\partial} \Omega^{(p,\,1)} \xrightarrow{\bar\partial} \dots \xrightarrow{\bar\partial} \Omega^{(p,\,n)} \longrightarrow 0, \tag{2.17}$$

called the *Dolbeault complex*. On the sheaves of locally defined forms, this sequence is exact, and one gets the theorem of Dolbeault that

$$\ker \bar\partial^{(p,\,q)} / \operatorname{range} \bar\partial^{(p,\,q-1)} \simeq \mathrm{H}^q\left(M,\, \mathcal{O}\left(\Omega^{(p,\,0)}\right)\right), \tag{2.18}$$

where the right hand side denotes the $q$-th cohomology group of the sheaf $\mathcal{O}\left(\Omega^{(p,\,0)}\right)$ of holomorphic $(p,\, 0)$-forms over $M$. See for instance Griffiths and Harris [33, p. 45]. If $M$ is compact, then the ellipticity of the complex yields that the spaces in the left hand side are finite-dimensional. We will mainly be interested in the case that $p = 0$.

A holomorphic vector bundle $L$ over $M$ is defined as a complex vector bundle over $M$ for which the retrivializations are given by elements of $\mathrm{GL}(l,\, \mathbf{C})$ which depend holomorphically on the base point. (Here $l = \dim_{\mathbf{C}} L_x$.) All the above remains valid for $L$-valued $(p,\, q)$-forms, that is, the sections of the vector bundle

$$\mathrm{T}_x^*\, M^{(p,\,q)} \otimes L_x, \quad x \in M.$$

In particular, we have the "twisted Dolbeault complex" defined by

$$\bar{\partial}^{(0,q)} : E^q \otimes L \to E^{q+1} \otimes L,$$

and the corresponding cohomology groups

$$\ker \bar{\partial}^{(0,q)} / \operatorname{range} \bar{\partial}^{(0,q-1)} \simeq \mathrm{H}^q(M, \mathcal{O}(L)), \tag{2.19}$$

the $q$-th cohomology group of the sheaf $\mathcal{O}(L)$ of holomorphic sections of $L$ over $M$.

If $M$ is a compact complex analytic manifold, then an important quantity is the *Riemann-Roch number*

$$\mathrm{RR}(M, L) := \sum_{q=0}^{n} (-1)^q \dim_{\mathbf{C}} \mathrm{H}^q(M, \mathcal{O}(L)). \tag{2.20}$$

More generally, if $\gamma$ is a complex analytic automorphism of $L$, then $\gamma$ acts on $\mathcal{O}(L)$, and one can define its *holomorphic Lefschetz number*

$$\chi(\gamma) \doteq \chi_{M,L}(\gamma) := \sum_{q=0}^{n} (-1)^q \operatorname{trace}_{\mathbf{C}} \gamma|_{\mathrm{H}^q(M,\mathcal{O}(L))}. \tag{2.21}$$

This *is* a generalization because $\chi_{M,L}(1) = \mathrm{RR}(M, L)$.

If $L$ is a holomorphic complex line bundle over $M$ for which $K^* \otimes L$ is positive, then Kodaira's vanishing theorem says that $\mathrm{H}^q(M, \mathcal{O}(L)) = 0$ for every $q > 0$, cf. (6.34). If the latter is the case, the holomomorphic Lefschetz number is equal to the trace of the action of $\gamma$ on the space $\mathrm{H}^0(M, \mathcal{O}(L))$ of all holomorphic sections of $L$ over $M$, and the Riemann-Roch number is equal to the dimension of that space.

## 2.2 The Dolbeault-Dirac Operator

In order to define adjoints, we now introduce Hermitian structures $h$ and $h^L$ in the tangent bundle $\mathrm{T}\,M$ of $M$ and the fibers of $L$, respectively. For each $x \in M$, $h_x$ is a complex-valued bilinear form on $\mathrm{T}_x\,M$, such that

$$h_x(v, v) > 0 \text{ if } v \in \mathrm{T}_x\,M,\ v \neq 0, \tag{2.22}$$

and

$$h_x\left(J_x(v),\, w\right) = i\, h_x(v,\, w) = -h_x\left(v,\, J_x(w)\right),\;\; v,\, w \in \mathrm{T}_x\, M. \tag{2.23}$$

Similarly with $h_x$ and $\mathrm{T}_x\, M$ replaced by $h_x^L$ and $L_x$, respectively.

It follows that the real part $\beta = \operatorname{Re} h$ of $h$ is a Riemannian structure in $M$, and $J_x$ is antisymmetric with respect to $\beta_x$. Furthermore, the imaginary part $\sigma = \operatorname{Im} h$ is a nowhere degenerate two-form in $M$, and $J_x$ is infinitesimally symplectic for $\sigma_x$. Finally,

$$\beta(v,\, w) = \sigma\left(Jv,\, w\right) \tag{2.24}$$

shows that choosing two of the three structures $J,\ \beta,\ \sigma$ determines the third.

In order to get a Hermitian structure in $E$, we begin by observing that

$$h_x : v \mapsto \left(w \mapsto h_x(v,\, w)\right) : \mathrm{T}_x\, M \longrightarrow \mathrm{T}_x^*\, M^{(0,\,1)} \tag{2.25}$$

is a complex linear isomorphism from $\mathrm{T}_x\, M$ onto $\mathrm{T}_x^*\, M^{(0,\,1)}$. Using this isomorphism, we transplant the Hermitian structure of $\mathrm{T}\, M$ to a Hermitian structure $h^{(0,\,1)}$ in $\mathrm{T}^*\, M^{(0,\,1)}$. That is, $h^{(0,\,1)}$ is determined by the condition that, if $e_j$, $1 \le j \le n$, is a unitary local frame in $\mathrm{T}\, M$ for $h$, then $\epsilon_j := h\, e_j$ forms a unitary local frame in $\mathrm{T}^*\, M^{(0,\,1)}$ for $h^{(0,\,1)}$. It is dual to the frame $e_j$, in the sense that $\langle e_j,\, \epsilon_k\rangle = \delta_{jk}$.

The Hermitian structure $h^{(0,\,q)}$ on $E^q = \mathrm{T}^*\, M^{(0,\,q)}$ can now be defined by the conditon that the $\epsilon_K$ form a unitary local frame in $E^q$, if for each strictly increasing sequence $K = (k_i)_{i=1}^q$ we write

$$\epsilon_K := \epsilon_{k_1} \wedge \epsilon_{k_2} \wedge \ldots \wedge \epsilon_{k_q}. \tag{2.26}$$

The Hermitian structure $h_E$ on the direct sum $E$ of the $E^q$ is defined by requiring the summands to be mutually orthogonal. And the Hermitian structure $h_{E \otimes L}$ on $E \otimes L$ by the condition that if $e_j$ and $l_k$ are unitary local frames in $E$ and $L$, respectively, then the $e_j \otimes l_k$ form a unitary local frame in $E \otimes L$.

The Hermitian $L^2$-inner product of two sections $u$ and $v$ of $E \otimes L$ is now defined as

$$(u, v) = \int_M h_{E \otimes L}(u, v) \, \mathrm{d}x, \tag{2.27}$$

where $\mathrm{d}x$ denotes the standard volume form in $M$, such that

$$\mathrm{d}x \, (e_1, J e_1, e_2, J e_2, \ldots, e_n, J e_n) = 1. \tag{2.28}$$

(Declaring $\mathrm{d}x$ to be positive determines the orientation of $M$.)

If $D : \Gamma (M, E \otimes L) \to \Gamma (M, E \otimes L)$ is a differential operator, then the (formal) *adjoint* $D^*$ of $D$ is defined by

$$(D(u), v) = (u, D^*(v)), \quad u, v \in \Gamma (M, E \otimes L), \tag{2.29}$$

where one of the $u$, $v$ is compactly supported. The adjective "formal" is added, if one wants to stress that no attention is being paid to questions of $L^2$-closure.

In complex local coordinates, and a local holomorphic trivialization of $L$, we get for a constant $\mathbf{C}^l$-valued $(0, q)$-form $\omega$ and real-valued linear form $\xi$ on $\mathbf{C}^n \simeq \mathbf{R}^{2n}$:

$$\mathrm{e}^{-i\langle z, \xi \rangle} \, \bar{\partial} \left( \mathrm{e}^{i\langle z, \xi \rangle} \, \omega \right) = i \, \xi^{(0,1)} \wedge \omega.$$

Because of its frequent occurrence, we will use the abbreviation

$$\mathrm{e}(\alpha) : \omega \mapsto \alpha \wedge \omega : E^q_x \otimes L_x \to E^{q+1}_x \otimes L_x \tag{2.30}$$

for the linear mapping of *taking the exterior product with* $\alpha \in \mathrm{T}^*_x M^{(0,1)}$ *from the left*. Using the invariance properties of principal symbols, we get that the principal symbol of the operator $\bar{\partial}$ at $\xi \in \mathrm{T}^*_x M$ is equal to

$$\sigma_{\bar{\partial}}(\xi) = i \, \mathrm{e} \left( \xi^{(0,1)} \right). \tag{2.31}$$

The principal symbol at $\xi \in \mathrm{T}^*_x M$ of the adjoint operator

$$\bar{\partial}^* : \Gamma \left( M, E^{q+1} \otimes L \right) \to \Gamma (M, E^q \otimes L) \tag{2.32}$$

is equal to the adjoint of (2.31) with respect to the given Hermitian structures. In order to compute this, we observe that for every $r$ and every pair of strictly increasing sequences $J = (j_i)_{i=1}^{q}$, $K = (k_i)_{i=1}^{q+1}$, we have $h_x(\epsilon_r \wedge \epsilon_J, \epsilon_K) = 0$, unless $|K| = \{r\} \cup |J|$, and in this case the inner product is equal to $(-1)^s$, where $s$ is equal to the number of $i$ such that $j_i < r$. Since the same answer is obtained for $h_x(\epsilon_J, \mathrm{i}(e_r)\,\epsilon_K)$, we conclude that the adjoint of $\mathrm{e}(\epsilon_r)$, the operator of taking exterior product with $\epsilon_r$ from the left, is equal to $\mathrm{i}(e_r)$, the operator of *taking inner product with* $e_r$. Writing $\xi_r = \langle e_r, \xi\rangle$, $\eta_r = \langle Je_r, \xi\rangle$, we get

$$\xi^{(0,1)} = \sum_{r=1}^{n} \tfrac{1}{2}(\xi_r + i\,\eta_r)\,\epsilon_r. \tag{2.33}$$

Moving over every $\epsilon_r$-term, and using that $h_x$ is complex linear in the first variable and complex antilinear in the second variable, we get

$$\sigma_{\bar\partial^*}(\xi)\,\omega = -\tfrac{i}{2}\sum_{r=1}^{n}(\xi_r - i\,\eta_r)\,\mathrm{i}(e_r)\,\omega.$$

Since the antilinearity of $\omega$ yields that

$$\mathrm{i}(Je_r)\,\omega = -i\,\mathrm{i}(e_r)\,\omega,$$

the result is [1]

$$\sigma_{\bar\partial^*}(\xi) = -\tfrac{i}{2}\,\mathrm{i}\left(\beta^{-1}\xi\right) : E_x^{q+1}\otimes L_x \to E_x^{q}\otimes L_x. \tag{2.34}$$

We now consider the operator $D = 2\left(\bar\partial + \bar\partial^*\right)$, acting on $\Gamma(M, E\otimes L)$, with principal symbol equal to $2(\sigma_{\bar\partial} + \sigma_{\bar\partial^*})$. Here the factor two has been inserted in order to cancel the halves in (2.31), cf. (2.33), and in (2.34).

---

[1] In [9, p. 138], [32, p. 184] and [33, p. 80] the more customary convention is used of taking the Hermitian inner product in $E_q$, such that $h_x(\epsilon_K, \epsilon_K) = 2^q$ for $\epsilon_K \in E^q$. As a consequence, (2.34) is equal to 1/2 times the formula for the symbol of $\bar\partial^*$, obtained in [32, p. 184] and [9, p. 138].

For any one-form $\alpha$ and vector $v$ we have $\mathrm{e}(\alpha)^2 = 0$, $\mathrm{i}(v)^2 = 0$, and

$$\mathrm{i}(v)\,(\alpha \wedge \omega) = \langle v,\, \alpha \rangle\, \omega - \alpha \wedge (\mathrm{i}(v)\, \omega)\,,$$

and therefore

$$(\mathrm{e}(\alpha) + \mathrm{i}(v))^2 = \langle v,\, \alpha \rangle.$$

It follows now from (2.31) and (2.34), that

$$\sigma_{D^2}(\xi) = \sigma_D(\xi)^2 = \langle {\beta_x}^{-1}\, \xi,\, \xi \rangle. \tag{2.35}$$

It is quite remarkable that the square of the principal symbol is a scalar, because the principal symbol itself is a linear transformation in $E_x \otimes L_x$, which is far from diagonal. It actually has no diagonal terms at all because it maps $E_x^q \otimes L_x$ to the sum of $E_x^{q+1} \otimes L_x$ and $E_x^{q-1} \otimes L_x$. In [9], an operator $D$ which satisfies (2.35) is called a *generalized Dirac operator*. Since the generalized Dirac operator $2\left(\bar{\partial} + \bar{\partial}^*\right)$ is defined in terms of the operator $\bar{\partial}$, which appears in the Dolbeault complex (2.17), I got into the habit of calling it the *Dolbeault-Dirac operator*.

From (2.35) we see that if $\xi \in \mathrm{T}_x\, M$ (real) and $\xi \neq 0$, then $\sigma_D(\xi)^2$ is invertible, and hence $\sigma_D(\xi)$ is invertible as well. That is, the operator $D$ is elliptic. If $M$ is compact, then this implies that the kernel of $D$ is finite-dimensional and the range of $D$ is equal to the orthogonal complement of the kernel of $D^* = D$. However $\bar{\partial}^2 = 0$ implies that $\left(\bar{\partial}^*\right)^2 = 0$ and hence

$$\tfrac{1}{4} D^2 = \bar{\partial}^*\, \bar{\partial} + \bar{\partial}\, \bar{\partial}^*$$

or

$$\tfrac{1}{4}\, (D\, u,\, D\, u) = \left(\bar{\partial}\, u,\, \bar{\partial}\, u\right) + \left(\bar{\partial}^*\, u,\, \bar{\partial}^*\, u\right).$$

So $D\, u = 0$ is equivalent to $\bar{\partial}\, u = 0$ and $\bar{\partial}^*\, u = 0$. Since the latter equation means that $u$ is in the orthogonal complement of the range of $\bar{\partial}$, we get from the Dolbeault theorem (2.18) that

$$\ker D|_{\Gamma(M,\, E^q \otimes L)} \simeq \mathrm{H}^q\,(M,\, \mathcal{O}(L))\,. \tag{2.36}$$

This allows us to translate (2.20) and (2.21) in terms of the kernel of $D = 2\left(\bar{\partial} + \bar{\partial}^*\right)$. Note that $D$ maps sections of $E^q \otimes L$ to sections of $(E^{q+1} \oplus E^{q-1}) \otimes L$, so we have the operators

$$D^+ = D|_{\Gamma(M,\,E^+ \otimes L)} : \Gamma\left(M,\, E^+ \otimes L\right) \to \Gamma\left(M,\, E^- \otimes L\right) \tag{2.37}$$

and

$$D^- = D|_{\Gamma(M,\,E^- \otimes L)} : \Gamma\left(M,\, E^- \otimes L\right) \to \Gamma\left(M,\, E^+ \otimes L\right). \tag{2.38}$$

Collecting the terms with the same sign in (2.20), we get

$$\begin{aligned} \mathrm{RR}(M,\, L) &= \dim_{\mathbf{C}} \ker D^+ - \dim_{\mathbf{C}} \ker D^- \\ &= \dim_{\mathbf{C}} \ker D^+ - \dim_{\mathbf{C}} \ker \left(D^+\right)^* = \operatorname{index} D^+, \end{aligned} \tag{2.39}$$

the *index* of the operator $D^+$. And doing the same in (2.21):

$$\chi(t) = \operatorname{trace}_{\mathbf{C}} \gamma|_{\ker D^+} - \operatorname{trace}_{\mathbf{C}} \gamma|_{\ker D^-}. \tag{2.40}$$

# Chapter 3

# Clifford Modules

## 3.1 The Non-Kähler Case

In general, if the manifold is not Kähler, then the Dolbeault-Dirac operator $D = 2\left(\bar{\partial} + \bar{\partial}^*\right)$ is *not* the most suitable one for getting explicit formulas for (2.39) and (2.40). For instance, if $M$ is a complex analytic manifold and $n = 2$, then Gilkey [29, Thm. 3.7] proved that the difference

$$\text{trace}_{\mathbf{C}}\, K_1^+(x) - \text{trace}_{\mathbf{C}}\, K_1^-(x)$$

of the traces of the coefficients of $t^{-1}$ in the asymptotic expansion (1.2) is equal to a universal constant times

$$\mathrm{d}\,\bar{\partial}\sigma = \partial\,\bar{\partial}\sigma = \partial\,\mathrm{d}\,\sigma. \tag{3.1}$$

Here the Kähler form $\sigma$, cf. (2.24), is viewed as a $(1,\,1)$-form. Moreover, the integrand in (1.3) is equal to a universal constant times the Laplacian applied to (3.1), plus terms involving at most second order derivatives of the metric. This follows from the formulas for $E_2$ and $E_4$ of Gilkey [30, p. 610], using that, in the computation of the supertrace, the linear contributions of the scalar curvature drop out. So the integrand in (1.3) involves fourth order derivatives of $\sigma$. I am grateful to Peter Gilkey for helping me with these references. Instead of the Dolbeault-Dirac operator, we shall therefore use the so-called spin-c Dirac operator. This is an operator with the same principal symbol as the Dolbeault-Dirac operator, and defined in terms of a carefully chosen connection in the bundle $E \otimes L$, see (5.14). The definition involves the bundle $\mathrm{C}\,(\mathrm{T}^*\, M)$ of Clifford algebras of the duals $\mathrm{T}_x^*\, M$ of the tangent spaces, which comes naturally with the principal symbol (2.35). Sections of $\mathrm{C}\,(\mathrm{T}^*\, M)$ act on sections of $E \otimes L$ by means of the so-called Clifford multiplication, and the crucial property of the connection which is used in the definition of the spin-c Dirac operator is that it yields a Leibniz rule of the form (3.39). Such a connection is called a *Clifford connection.* In this chapter we will discuss the Clifford algebras. The Clifford algebra $\mathrm{C}(2n)$ contains as a natural subgroup the spin group $\mathrm{Spin}(2n)$, which is related to the rotation group. And the complexification of $\mathrm{C}(2n)$ contains the slightly larger spin-c group $\mathrm{Spin}^{\mathrm{c}}(2n)$, which contains the unitary group $\mathrm{U}(n)$ in a natural way. These groups will be introduced the next chapter, and will be used in the definition of the connection in $E \otimes L$ and the spin-c Dirac operator in Chapter 5. In the condition (3.39) for a Clifford connection, there appears a connection in the bundle $\mathrm{C}\,(\mathrm{T}^*\, M)$ of Clifford algebras, which is defined in terms of the so-called *Levi-Civita connection* of the Riemannian structure $\beta$. This is the unique connection in $\mathrm{T}\, M$, which leaves $\beta$ invariant and is *torsion-free.* Recall that for a general linear connection in the tangent bundle, with the corresponding covariant differentiation $\nabla$ of vector fields in $M$, the *torsion*$T_x$ at $x \in M$ is the bilinear mapping from $\mathrm{T}_x\, M \times \mathrm{T}_x\, M$ to $\mathrm{T}_x\, M$, defined by

$$T(v,\, w) = \nabla_v w - \nabla_w v - [v,\, w], \tag{3.2}$$

for any vector fields $v,\ w$ in $M$. So the connection is torsion-free, if and only if

$$[v,\, w] = \nabla_v w - \nabla_w v, \tag{3.3}$$

expressing the Lie bracket of vector fields in terms of the covariant derivatives. Before we continue, we warn that, in general, the Levi-Civita connection does not leave the almost complex structure invariant. If it would, then for any pair of vector fields $v$, $w$ in $M$, each of the expressions

$$\begin{aligned} &\nabla_v (Jw) - J\,\nabla_v w, \\ &-\nabla_w (Jv) + J\,\nabla_w v, \\ &-\nabla_{Jw} v - J\,\nabla_{Jw} (Jv)\,, \\ &\nabla_{Jv} w + J\,\nabla_{Jv}\,(Jw)\,, \end{aligned}$$

is equal to zero. Summing, one gets that

$$[v,\, Jw] + [Jv,\, w] + J\,[Jv,\, Jw] - J\,[v,\, w] = 0,$$

which is equivalent to $[J,\, J] = 0$, or $M$ is complex analytic. See the comments after the definition (2.15) of $[J,\, J]$. A similar computation shows that $\mathrm{d}\sigma = 0$, so $(M,\, J,\, h)$ is a Kähler manifold, and this conversely implies that the Levi-Civita connection leaves $J$ invariant, cf. [9, Proposition 3.66]. So the warning is that *in the non-Kähler case, a connection in $T\,M$, which leaves both $\beta$ and $J$ invariant, and therefore also $h$ and $\sigma$, necessarily has nonzero torsion.* The fact that in the non-Kähler case the Levi-Civita connection does not leave the almost complex structure invariant is the reason for replacing the Dolbeault-Dirac operator by the spin-c Dirac operator. For the spin-c Dirac operator, we will get that the difference of the traces of $K^{+}(t,\, x,\, x)$ and $K^{-}(t,\, x,\, x)$ converges as $t \downarrow 0$, and the limit can be determined explicitly (Theorem 11.1). A similar result holds in the presence of an automorphism $\gamma$ (Theorem 12.1). Since the spin-c Dirac operator has the same principal symbol as the Dolbeault-Dirac operator, this leads, if $M$ is compact, to a determination of the Riemann-Roch number and the virtual character as integrals over $M$ and $M^{\gamma}$ of these respective limits. In comparison with the Dolbeault-Dirac operator, the drawback of the spin-c Dirac operator is that in the case of a complex analytic manifold which is not Kähler, its kernel does not consist of holomorphic sections. Also, in general the square of the spin-c Dirac operator does not preserve the degrees of the differential forms (it only preserves the degree modulo two), so it cannot be used to get information about sections in the kernel of $D$ which have a given degree as a differential form.

## 3.2 The Clifford Algebra

The point of departure for the definition of the spin-c Dirac operator is the formula (2.35) for the principal symbol, which leads naturally to a Clifford module structure on the fibers $E_x$ of the bundle $E$. We shortly repeat the definitions and some of the properties which we will need. We refer to [9, Chapter 3] and Lawson and Michelsohn [51, Ch. 1] for some more details. If $V$ is a real vector space, provided with an inner product $Q$, then the *Clifford algebra* $\mathrm{C}(V) = \mathrm{C}\,(V,\, Q)$ is the algebra over $\mathbf{R}$, generated by $V$, and with the relation

$$v \cdot v = -Q\,(v,\, v)\,, \quad \text{for each } v \in V. \tag{3.4}$$

More precisely, $\mathrm{C}(V)$ is defined as the quotient of the tensor algebra $\bigotimes V$ of $V$ (with unit) by the two-sided ideal $I$, generated by the set of the elements $v \otimes v + Q\,(v,\, v)$, with $v \in V$. Note that (3.4) is equivalent to

$$v \cdot w + w \cdot v = -2Q\,(v,\, w)\,, \quad v,\, w \in V. \tag{3.5}$$

If one works over the complex numbers, then the rank of $Q$ determines the Clifford algebras up to isomorphism, whereas in the real case the signature also matters. We will follow the original convention of Clifford [17] (whose point is that $V$ can have any dimension) and take $Q$ to be positive definite. In Dirac's papers [18], $V$ is four-dimensional and $Q$ has the Lorentz signature, whereas in Brauer and Weyl [15], $Q$ is chosen to be negative definite (and $\dim V$ arbitrary). The map $v \mapsto v + I : V \to \mathrm{C}(V)$ is an injective linear mapping from $V$ to $\mathrm{C}(V)$, which will be used in order to identify $V$ with a linear subspace of $\mathrm{C}(V)$. One also has $1 \notin V$. The grading of the tensor algebra $\bigotimes V$ passes to a filtration of $\mathrm{C}(V)$: $\mathrm{C}^{\leq q}(V)$ is the linear subspace of $\mathrm{C}(V)$ generated by the

$$v_1 \otimes v_2 \otimes \ldots \otimes v_k + I, \tag{3.6}$$

with $v_j \in V$ and $k \leq q$. Note that $\mathrm{C}^{\leq 1}(V) = \mathbf{R} \oplus V$, and that

$$\mathrm{C}^{\leq q}(V) \cdot \mathrm{C}^{\leq r}(V) \subset \mathrm{C}^{\leq (q+r)}(V). \tag{3.7}$$

The grading of $\bigotimes V$ also leads to a grading modulo two in $\mathrm{C}(V)$; the vector space in $\mathrm{C}(V)$ generated by the (3.6) with $k$ even and with $k$ odd will be denoted by $\mathrm{C}^{+}(V)$ and $\mathrm{C}^{-}(V)$, respectively. We have the following properties:

$$\mathrm{C}(V) = \mathrm{C}^{+}(V) \oplus \mathrm{C}^{-}(V), \tag{3.8}$$

$$\mathrm{C}^{+}(V) \cdot \mathrm{C}^{+}(V) \subset \mathrm{C}^{+}(V), \tag{3.9}$$

$$\mathrm{C}^{+}(V) \cdot \mathrm{C}^{-}(V) \subset \mathrm{C}^{-}(V), \tag{3.10}$$

$$\mathrm{C}^{-}(V) \cdot \mathrm{C}^{+}(V) \subset \mathrm{C}^{-}(V), \tag{3.11}$$

$$\mathrm{C}^{-}(V) \cdot \mathrm{C}^{-}(V) \subset \mathrm{C}^{+}(V). \tag{3.12}$$

This means that $\mathrm{C}(V)$ is a *superalgebra*; *not* supercommutative as soon as $Q \neq 0$. We also have $\mathrm{C}^{\leq 1}(V) \cap \mathrm{C}^{-}(V) = V$ and $\mathrm{C}^{\leq 1}(V) \cap \mathrm{C}^{+}(V) = \mathbf{R}$. In particular, $\mathrm{C}^{+}(V)$ is a subalgebra of $\mathrm{C}(V)$, and the invertible elements in it form a Lie group, which will be denoted by $\mathrm{C}^{+}(V)^{\times}$. Clifford [17] remarks that if $V = \mathbf{R}^3$ and $Q$ is the standard inner product in $\mathbf{R}^3$, then $\mathrm{C}^{+}(V)$ is the algebra of the quaternions.The basic property of the algebra $\mathrm{C}(V)$, which actually determines it up to isomorphism, is the following. Let $A$ be any real algebra with unit, and let $\phi$ be a real linear map from $V$ to $A$ such that

$$\phi(v)^2 = -Q\,(v,\, v)\,, \quad \text{for each } v \in V. \tag{3.13}$$

Then there is a unique extension of $\phi$ to a homomorphism from $\mathrm{C}(V)$ to $A$, which also will be denoted by $\phi$. If $A$ happens to be a complex algebra, then the map $\phi : \mathrm{C}(V) \to A$ has a unique extension to a complex linear map from $\mathrm{C}(V) \otimes \mathbf{C}$ to $A$, which automatically is a homomorphism of complex algebras and again will be denoted by $\phi$.

As a first application of this principle, consider the linear mapping $\mathrm{c}_\Lambda$, which assigns to each $\xi \in V$ the endomorphism $\mathrm{c}_\Lambda(\xi)$ in the exterior algebra $\Lambda V$ of $V$, defined by

$$\mathrm{c}_\Lambda(\xi)(\omega) = \xi \wedge \omega - \mathrm{i}\,(Q\,\xi)\,\omega. \tag{3.14}$$

This induces an algebra homomorphism

$$\mathrm{c}_\Lambda : \mathrm{C}(V) \to \mathrm{End}\,(\Lambda V)\,.$$

Then the map

$$\sigma_\Lambda : a \mapsto \mathrm{c}_\Lambda(a)(1) : \mathrm{C}(V) \to \Lambda V \tag{3.15}$$

has the property that for any $v_1,\, v_2,\, \ldots,\, v_k \in V$:

$$\sigma_\Lambda\,(v_1 \cdot v_2 \cdot\, \ldots \cdot v_k) = v_1 \wedge v_2 \wedge \ldots \wedge v_k \ \text{ modulo } \bigoplus_{l \geq 1} \Lambda^{k-2l} V. \tag{3.16}$$

It follows that $\sigma_\Lambda$ is a linear isomorphism, and preserves the filtration and the grading modulo 2. The formula

$$\sigma_\Lambda(\omega) \cdot_{_Q} \sigma_\Lambda(\nu) := \sigma_\Lambda\,(\omega \cdot \nu)$$

defines a $Q$-dependent product

$$(\alpha,\, \beta) \mapsto \alpha \cdot_{_Q} \beta$$

in $\Lambda V$, which can be viewed as a "lower order perturbation" of the wedge product, because if $\alpha \in \Lambda^a V$ and $\beta \in \Lambda^b V$, then

$$\alpha \cdot_{_Q} \beta - \alpha \wedge \beta \in \Lambda^{a+b-2} V.$$

If $a^{\pm}$, $b^{\pm} \in \mathrm{C}^{\pm}(V)$, then one defines the *supercommutator* of $a = a^+ + a^-$ and $b = b^+ + b^-$ by

$$[a, b]_{\mathrm{s}} := \left[a^+, b^+\right] + \left[a^+, b^-\right] + \left[a^-, b^+\right] + a^- b^- + b^- a^- . \qquad (3.17)$$

If $v \in V$, then the operator

$$\mathrm{ad_s}(v) : a \mapsto [v, a]_{\mathrm{s}}$$

in $\mathrm{C}(V)$ is a *superderivation* in the sense that

$$\mathrm{ad_s}(v)(a \cdot b) = \mathrm{ad_s}(v)(a) \cdot b \pm a \cdot \mathrm{ad_s}(v)(b), \qquad (3.18)$$

if $a \in \mathrm{C}^{\pm}(V)$. Since $\mathrm{i}(Qv)$ is a superderivation in $\Lambda V$, it follows that the formula

$$\sigma_{\Lambda}\left(\mathrm{ad_s}(v)(a)\right) = -2\ \mathrm{i}(Qv)\, \sigma_{\Lambda}(a), \qquad (3.19)$$

which is valid for $a \in V$ by the definition of $\mathrm{C}(V)$, extends its validity to arbitrary $a \in \mathrm{C}(V)$. In particular, $\mathrm{ad_s}(v)$ maps $\mathrm{C}^{\leq k}(V)$ into $\mathrm{C}^{\leq (k-1)}(V)$. This in turn can be used to prove that the algebra $\mathrm{C}(V)$ is *simple*, which means that if $I$ is an nonzero two-sided ideal in $\mathrm{C}(V)$, then $I = \mathrm{C}(V)$. Indeed, let $a \in I$, $a \neq 0$, and let $k$ be the order of $a$, that is, $a \in \mathrm{C}^{\leq k}(V)$ but $a \notin \mathrm{C}^{\leq (k-1)}(V)$. Then we see from (3.19) that we can find $v_1, \ldots, v_k \in V$, such that the element

$$t = \mathrm{ad_s}(v_1) \circ \ldots \circ \mathrm{ad_s}(v_k)(a) \in I$$

is nonzero. But $t \in \mathrm{C}^{\leq 0}(V) = \mathbf{R}$, so the fact that $I$ is an ideal and $t \in I$ now implies that $I = \mathrm{C}(V)$.

We now assume that $V = W^*$, the dual of a real vector space $W$ of dimension $2n$, and that $W$ is provided with a complex structure $J$ and a Hermitian bilinear form $h$. Then $\beta = \mathrm{Re}\, h$ is an inner product in $W$ and we take $Q = \beta^{-1}$ equal to the corresponding inner product in $V = W^*$. Finally, the natural complex structure in $V$ is given by

$$J := -J' : V \to V, \qquad (3.20)$$

if $J'$ denotes the real transposed of $J$. In this notation, (2.4) reads:

$$\xi^{(0,1)} = \tfrac{1}{2}\,(\xi - i\,J\xi)\,. \tag{3.21}$$

In this case, the complexification $V \otimes \mathbf{C}$ is equal to the direct sum of the complex linear subspace

$$V^{(1,0)} := \{\xi + i\,J\xi \mid \xi \in V\} \tag{3.22}$$

of forms on $W$ which are complex-linear with respect to $J$, and the space

$$V^{(0,1)} := \{\xi - i\,J\xi \mid \xi \in V\} \tag{3.23}$$

of forms on $W$ which are complex-antilinear with respect to $J$. Write $E = \Lambda V^{(0,1)}$ and define, for each $\xi \in V$, the complex linear transformation $\mathrm{c}(\xi)$ in $E$ by

$$\mathrm{c}(\xi)(\omega) := (\xi - i\,J\xi) \wedge \omega - \mathrm{i}\,(Q\,\xi)\,\omega. \tag{3.24}$$

A computation as in (2.35) yields that

$$\mathrm{c}(\xi)^2 = -Q\,(\xi,\,\xi) \cdot 1, \tag{3.25}$$

so $\mathrm{c} : V \to \mathrm{End}(E)$ has a unique extension to a homomorphism of complex algebras

$$\mathrm{c} : \mathrm{C}(V) \otimes \mathbf{C} \to \mathrm{End}(E). \tag{3.26}$$

The symbol $\mathrm{c}(a)$ stands for *multiplication in $E$ by the element $a$ of the Clifford algebra* $\mathrm{C}(V)$. In this way the homomorphism (3.26) turns $E$ into a $\mathrm{C}(V)$-module, a *Clifford module.*

**Lemma 3.1** *Let $E$ be the exterior algebra of $V^{(0,1)}$. Then the homomorphism (3.26), which extends (3.24), is an isomorphism of the complex algebra* $\mathrm{C}(V) \otimes \mathbf{C}$ *onto* $\mathrm{End}(E)$.

**Proof** The argument that the algebra $\mathrm{C}(V)$ is simple works the same for the complexified algebra $\mathrm{C}(V) \otimes \mathbf{C}$. Since the homomorphism c is clearly nonzero, its kernel is a two-sided ideal $I$ in $\mathrm{C}(V) \otimes \mathbf{C}$, so $I = 0$, or c is injective. On the other hand, $\mathrm{C}(V) \otimes \mathbf{C}$ and $\mathrm{End}(E)$ have the same dimension over $\mathbf{C}$, namely $2^{2n} = (2^n)^2$. The conclusion is that the linear mapping c is surjective as well. □

One can view Lemma 3.1 as a structure theorem for the Clifford algebra $\mathrm{C}(V)$, but we will use it in order to transfer Clifford algebra structures, such as its filtration, to $\mathrm{End}(E)$. With the notation of Chapter 2, we will apply this to $V = \mathrm{T}_x^* M$, $W = \mathrm{T}_x M$, $J = J_x$, $Q = {\beta_x}^{-1}$, and $E = E_x$. Comparing the sum of (2.31) and (2.34) with (3.24), we see that $\sigma_D = i\ \mathrm{c}$. This is the reason for our choice of c in (3.24)[1]. If $E^+$ and $E^-$ denote the sum in $E$ of the spaces $\Lambda^q V^{(0,\,1)}$ with $q$ even and with $q$ odd, respectively, then we have, analogously to (3.8) — (3.12):

$$E = E^+ \oplus E^-, \tag{3.27}$$

$$\left(\mathrm{C}^+(V) \otimes \mathbf{C}\right) \cdot_{\mathrm{c}} E^+ \subset E^+, \tag{3.28}$$

$$\left(\mathrm{C}^+(V) \otimes \mathbf{C}\right) \cdot_{\mathrm{c}} E^- \subset E^-, \tag{3.29}$$

$$\left(\mathrm{C}^-(V) \otimes \mathbf{C}\right) \cdot_{\mathrm{c}} E^+ \subset E^-, \tag{3.30}$$

$$\left(\mathrm{C}^-(V) \otimes \mathbf{C}\right) \cdot_{\mathrm{c}} E^- \subset E^+. \tag{3.31}$$

In particular, $E^+$ and $E^-$ are $\mathrm{C}^+(V) \otimes \mathbf{C}$-modules.

## 3.3 The Supertrace

We interrupt the definition of the operators $D^{\pm}$, for a preview on the proof of the local formula for (2.40). In the computations, we will meet the *supertrace*

$$\mathrm{str}_{\mathbf{C}}\, A := \mathrm{trace}_{\mathbf{C}}\, A^{++} - \mathrm{trace}_{\mathbf{C}}\, A^{--} \tag{3.32}$$

[1] In [9, p. 110], the choice for $\mathrm{c}(\xi)$ is $\sqrt{2}$ times the difference of $\mathrm{e}\left(\xi^{(0,\,1)}\right)$ and $\mathrm{i}\,(Q\,\xi)$, which has the same square.

of endomorphisms $A$ of $E$. Here we denoted by $A^{jk}$ the restriction of $A$ to $E^k$, followed by the projection from $E$ to $E^j$ along the complementary subspace. If $A = \mathrm{c}(a)$, then $\mathrm{c}(a^+)$ and $\mathrm{c}(a^-)$ corresponds with the diagonal and the antidiagonal part of the block decomposition of $A$, respectively. If $A = \mathrm{c}(a)$ and $B = \mathrm{c}(b)$, then with the abbreviation $\mathrm{t} = \mathrm{trace}_{\mathbf{C}}$:

$$\begin{aligned}
&\mathrm{str}_{\mathbf{C}}\, \mathrm{c}(a^- b^- + b^- a^-) \\
&= \mathrm{t}\left(A^{+-} B^{-+} + B^{+-} A^{-+}\right) - \mathrm{t}\left(A^{-+} B^{+-} + B^{-+} A^{+-}\right) \\
&= \left(\mathrm{t}(A^{+-} B^{-+}) - \mathrm{t}(B^{-+} A^{+-})\right) + \left(\mathrm{t}(B^{+-} A^{-+}) - \mathrm{t}(A^{-+} B^{+-})\right) = 0.
\end{aligned}$$

Here the last identity can be seen by identifying $E^+$ and $E^-$ by means of an isomorphism. So, with the supercommutator of (3.17), we have

$$\mathrm{str}_{\mathbf{C}}\ \mathrm{c}\left([a,\, b]_{\mathrm{s}}\right) = 0$$

for all $a,\, b \in \mathrm{C}(V) \otimes \mathbf{C}$. Using (3.19), we see readily that

$$\sigma_\Lambda\left([\mathrm{C}(V),\, \mathrm{C}(V)]_{\mathrm{s}}\right) = \Lambda^{\leq (2n-1)} V := \bigoplus_{j=0}^{2n-1} \Lambda^j V.$$

It follows that $\mathrm{str}_{\mathbf{C}}(A) = 0$ if and only if $\sigma_\Lambda(A)$ belongs to the codimension one linear subspace $\Lambda^{\leq(2n-1)}V$ of $\Lambda V$. In other words, the supertrace of $A$ only depends on the component in $\Lambda^{2n}V$, the *volume part*, of $\sigma_\Lambda\left(\mathrm{c}^{-1}(A)\right)$ in $\Lambda V$. The inner product in $V$, together with the orientation of $V$, induces an isomorphism of $\Lambda^{2n}V$ with $\mathbf{R}$. Let us write $\mathrm{vol}(\omega) \in \mathbf{R}$ for the real number, which in this way is assigned to the volume part of $\omega \in \Lambda V$. In [9], the linear form vol on $\Lambda V$ is called *Berezin integration*, see also [8, pp. 52, 53]. After tensoring with $\mathbf{C}$, one now gets[2]

$$\mathrm{str}_{\mathbf{C}}\, A = \left(\tfrac{2}{i}\right)^n \mathrm{vol}\left(\sigma_\Lambda\ \mathrm{c}^{-1}(A)\right), \quad A \in \mathrm{End}(E). \tag{3.33}$$

[2]This is the formula in [27, Th. 1.8] or [9, Prop. 3.21], with $n$ replaced by $2n$, because in our case the real dimension of $V$ is equal to $2n$.

A version of (3.33) has been used by Patodi [64] in his proof of Hirzebruch's Riemann-Roch theorem. Another proof of (3.33) will be given in Section 4.3. The identity (3.33) will be one of the keys in the proof of the local formula for (2.40). See Chapter 11. It can be viewed as a transfer of the computation of the supertrace in the endomorphism algebra of $E$, to a determination of the volume part in the exterior algebra of $V$. This leads in a natural way to an expression of the integrands, in (7.4) and (7.6), in terms of differential forms.

## 3.4 The Clifford Bundle

In our application of the Clifford multiplication, we will take, for each $x \in M$, the vector space $V$ equal to $\mathrm{T}_x^* M$, so $W = \mathrm{T}_x M$, $J = J_x$ and $Q = {\beta_x}^{-1}$. In this way we get the Clifford algebra $\mathrm{C}\left(\mathrm{T}_x^* M\right)$ and the Clifford multiplication

$$\mathrm{c}_x : \mathrm{C}\left(\mathrm{T}_x^* M\right) \otimes \mathbf{C} \xrightarrow{\sim} \mathrm{End}\left(E_x\right). \tag{3.34}$$

The $\mathrm{C}\left(\mathrm{T}_x^* M\right)$, $x \in M$, form a smooth bundle $\mathrm{C}\left(\mathrm{T}^* M\right)$ over $M$, and for each section $a$ and $\omega$ of $\mathrm{C}\left(\mathrm{T}^* M\right)$ and $E$, respectively, we get the section $\mathrm{c}(a)\left(\omega\right)$ of $E$, defined by

$$\left(\mathrm{c}(a)\left(\omega\right)\right)_x := \mathrm{c}_x\left(a_x\right)\left(\omega_x\right), \quad x \in M. \tag{3.35}$$

On the other hand, the Clifford algebra of $\mathbf{R}^{2n}$, provided with the standard inner product, is denoted by $\mathrm{C}\left(2n\right)$. Each $A \in \mathrm{SO}(2n)$ induces an automorphism $A_{\mathrm{C}}$ of $\mathrm{C}\left(2n\right)$.For each orthonormal frame $f_x$ in $\mathrm{T}_x M$, regarded as a linear isomorphism from $\mathbf{R}^n$ onto $\mathrm{T}_x M$, the inverse of the transposed linear mapping ${f_x}'$ is a linear isomorphism from $\left(\mathbf{R}^n\right)^*$ onto $\mathrm{T}_x^* M$. Identifying $\left(\mathbf{R}^n\right)^*$ with $\mathbf{R}^n$ and regarding $\mathrm{T}_x^* M$ as a subset of the algebra $\mathrm{C}\left(\mathrm{T}_x M\right)$, we get

$$\left(\left({f_x}'\right)^{-1}\left(\xi\right)\right)^2 = \|\xi\|^2,$$

so $\left(f_x{}'\right)^{-1}$ extends uniquely to an isomorphism

$$f_{x,\mathrm{C}} : \mathrm{C}\,(2n) \xrightarrow{\sim} \mathrm{C}\,(\mathrm{T}_x^*\, M)\,.$$

For every $A \in \mathrm{SO}\,(2n)$, we have

$$f_{x,\mathrm{C}} \circ A_{\mathrm{C}} = (f_x \circ A)_{\mathrm{C}}\,,$$

which justifies dropping the subscripts C from the notation.The bundle $\mathrm{SOF}\, M$ of *oriented orthonormal frames* in $\mathrm{T}\, M$ is a principal $\mathrm{SO}\,(2n)$-bundle, if we let $A \in \mathrm{SO}\,(2n)$ act on it by $f_x \mapsto f_x \circ A^{-1}$. We see that the fibers of the mapping

$$(f_x,\, a) \mapsto f_x(a) : \mathrm{SOF}\, M \times \mathrm{C}\,(2n) \to \mathrm{C}\,(\mathrm{T}^*\, M)$$

are equal to the orbits of the $\mathrm{SO}\,(2n)$-action on the Cartesian product, if the action of $A \in \mathrm{SO}\,(2n)$ is given by

$$(f,\, a) \to \left(f \circ A^{-1},\, A_{\mathrm{C}}(a)\right). \tag{3.36}$$

In this way $\mathrm{C}\,(\mathrm{T}^*\, M)$ is identified with the associated bundle

$$\mathrm{C}\,(\mathrm{T}^*\, M) = \mathrm{SOF}\, M \times_{\mathrm{SO}(2n)} \mathrm{C}\,(2n)\,. \tag{3.37}$$

The identification (3.37) allows us to identify sections of $\mathrm{C}\,(\mathrm{T}^*\, M)$ with $\mathrm{SO}\,(2n)$-equivariant mappings from $\mathrm{SOF}\, M$ to $\mathrm{C}\,(2n)$. An $\mathrm{SO}\,(2n)$-invariant connection in $\mathrm{SOF}\, M$, which is the same as a linear connection in $\mathrm{T}\, M$ which leaves the Riemannian structure $\beta$ invariant, then leads to the covariant differentiation $\nabla$ of sections of $\mathrm{C}\,(\mathrm{T}^*\, M)$, defined by

$$\nabla_v a = v_{\mathrm{hor}} a. \tag{3.38}$$

Here $v$ denotes a vector field in $M$, $v_{\mathrm{hor}}$ is its horizontal lift in $\mathrm{SOF}\, M$, and in the right hand side of (3.38), $a$ is viewed as a $\mathrm{C}\,(2n)$-valued function on $\mathrm{SOF}\, M$. The connection in $\mathrm{T}\, M$ used in the right hand side of (3.38) is the Levi-Civita connection, the torsion-free one which leaves $\beta$ invariant. The right hand side of (3.38) defines a section of $\mathrm{C}\,(\mathrm{T}^*\,M)$ because it is $\mathrm{SO}\,(2n)$-equivariant. Our next task is to provide the bundle $E$ with a *Clifford connection*, a linear connection $\nabla = \nabla^E$, such that we have the Leibniz rule

$$\nabla^E_v\left(\mathrm{c}(a)\,(\omega)\right) = \mathrm{c}\left(\nabla_v a\right)\,(\omega) + \mathrm{c}(a)\,\left(\nabla^E_v\,\omega\right), \tag{3.39}$$

for each vector field $v$ in $M$ and sections $a$ and $\omega$ of $\mathrm{C}\,(\mathrm{T}^*\,M)$ and $E$, respectively. It is sufficient to have this for sections $a$ of $\mathrm{T}^*\,M$, that is, for one-forms in $M$. The complication however is that in (3.39) the product is the Clifford multiplication, defined by (3.24), with $Q = \beta^{-1}$. Another way of phrasing (3.39) is that *the Clifford multiplication* c *is covariantly constant.* A first attempt to do this is to introduce the bundle $\mathrm{UF}\,M$ of *unitary frames* with respect to the almost complex structure $J$. The unitary frame $e_x$ gives rise to the identification

$$z \mapsto \sum_{r=1}^{n} \mathrm{Re}\, z_r\, e_r + \mathrm{Im}\, z_r\, J_x\, e_r : \mathbf{C}^n \xrightarrow{\sim} \mathrm{T}_x\, M,$$

which we also denote by $e_x$. Note that $e_x \circ J = J_x \circ e_x$, if $J$ denotes the multiplication by $i$ in $\mathbf{C}^n$. The inverse of the pull-back $e_x^*$ by $e_x$ defines a complex linear isomorphism of exterior algebras

$$e_{x,\Lambda} := \left(e_x^*\right)^{-1} : \mathrm{E}(n) \xrightarrow{\sim} E_x,$$

if $\mathrm{E}(n)$ denotes the algebra of complex-antilinear forms on $\mathbf{C}^n$. Each linear mapping $A : V \to V$ which commutes with $J$ induces a complex-linear transformation

$$\Lambda(A) := (A')^* : E \to E. \tag{3.40}$$

Here $A' : W \to W$ denotes the transposed of $A$, a real linear mapping which commutes with $J$. So the pull-back $(A')^*$ by $A'$ maps complex antilinear forms on $W$ to complex antilinear forms on $W$. In this way, it defines a homomorphism from $E$ to itself. In turn, the mapping $A \mapsto \Lambda(A)$ is a homomorphism of algebras

$$\Lambda : \mathrm{End}\,(V,\, J) \to \mathrm{End}(E). \tag{3.41}$$

Although it makes the notation a bit more cumbersome, we have chosen to distinguish $\Lambda(A)$ from $A$ in the notation. In the case that $W = \mathbf{C}^n$, we have $\Lambda(A) \in \mathrm{End}\,(\mathrm{E}(n))$ and

$$(e_x \circ A')_\Lambda \circ \Lambda(A) = e_{x,\,\Lambda}.$$

So the mapping

$$(e_x,\, \omega) \mapsto e_{x,\,\Lambda}(\omega) : \mathrm{UF}\, M \times \mathrm{E}(n) \to E$$

leads to the identification

$$E = \mathrm{UF}\, M \times_{\mathrm{U}(n)} \mathrm{E}(n) \tag{3.42}$$

of $E$ with a bundle associated to the unitary frame bundle. Here the action of $A \in \mathrm{U}(n)$ on $\mathrm{UF}\, M \times \mathrm{E}(n)$ is given by

$$(e,\, \omega) \mapsto \left(e \circ A^{-1},\, \Lambda(A)\, \omega\right). \tag{3.43}$$

The unitary frame bundle $\mathrm{UF}\, M$ is identified with a subbundle of the oriented orthogonal frame bundle $\mathrm{SOF}\, M$ by assigning to the unitary frame $e = (e_1,\, e_2,\, \ldots,\, e_n)$ the oriented orthonormal frame

$$f = f(e) = (e_1,\, Je_1,\, e_2,\, Je_2,\, \ldots,\, e_n,\, Je_n). \tag{3.44}$$

Let us assume for a moment that the horizontal spaces for the connection (the Levi-Civita one) in $\mathrm{SOF}\, M$ are tangent to $\mathrm{UF}\, M$, so that these define a connection in $\mathrm{UF}\, M$. Then we use (3.42) to identify the sections of $E$ with $\mathrm{U}(n)$-equivariant mappings from $\mathrm{UF}\, M$ to $\Lambda\mathbf{C}^n$ and to define the covariant derivative with respect to the vector field $v$ by

$$\nabla_v^E \omega = v_{\mathrm{hor}}\omega. \tag{3.45}$$

The Leibniz rule (3.39) then follows from (3.38), (3.45) and the Leibniz rule

$$v_{\text{hor}} \left( \mathrm{c}(a) \, (\omega) \right) = \mathrm{c} \left( v_{\text{hor}} a \right) \, (\omega) + \mathrm{c}(a) \, \left( v_{\text{hor}} \omega \right) . \tag{3.46}$$

Note that in (3.46) the Clifford multiplication

$$\mathrm{c} : \mathrm{C} \, (2n) \rightarrow \mathrm{End} \, (\mathrm{E}(n))$$

does not depend on the point $x$ in $M$. Now the condition that the horizontal spaces are tangent to the unitary frame bundle just means that the Levi-Civita connection leaves the almost complex structure invariant, which, as we have seen before, is equivalent to the condition that $(M, \ J, \ h)$ is a Kähler manifold. In the non-Kähler case, we will follow Kawasaki [45, p. 156], and exhibit $E$ as an associated vector bundle with a bigger structure groupSpin$^{\mathrm{c}}$ $(2n)$, which contains $\mathrm{U}(n)$ and projects to $\mathrm{SO}(2n)$. This associated vector bundle then can be provided with a Clifford connection, which in turn is used in the definition in Chapter 5 of the spin-c Dirac operator. In the Kähler case, the spin-c Dirac operator is equal to the Dolbeault-Dirac operator, but in the non-Kähler case it is different.

# Chapter 4

# The Spin Group and the Spin-c Group

## 4.1 The Spin Group

We begin the definition of the group $\mathrm{Spin}^{\mathrm{c}}(V)$, introduced by Atiyah-Bott-Shapiro [4], with the definition of the slightly smaller *spin group* $\mathrm{Spin}(V)$.

If $\xi,\,\eta,\,\zeta \in V \subset \mathrm{C}(V)$, then

$$\xi \cdot \eta \cdot \zeta = -\xi \cdot \zeta \cdot \eta - 2Q\,(\eta,\,\zeta)\;\xi = \zeta \cdot \xi \cdot \eta + 2Q\,(\xi,\,\zeta)\;\eta - 2Q\,(\eta,\,\zeta)\;\xi.$$

So, if we write

$$a = \tfrac{1}{2}\,(\xi \cdot \eta - \eta \cdot \xi) \in \mathrm{C}^{\leq 2}(V) \cap \mathrm{C}^{+}(V),$$

then

$$\tau'(a)(\zeta) := a \cdot \zeta - \zeta \cdot a = 2\,Q\,(\xi,\,\zeta)\;\eta - 2\,Q\,(\eta,\,\zeta)\;\xi. \tag{4.1}$$

This shows that $\tau'(a)$ leaves $V$ invariant, and in $V$ acts as an antisymmetric operator, an element of the Lie algebra $\mathfrak{so}(V)$ of the rotation group $\mathrm{SO}(V)$ in $V$ with respect to the inner product $Q$. The linear subspace $\mathfrak{spin}(V)$ of $\mathrm{C}^{\leq 2}(V) \cap \mathrm{C}^{+}(V)$, spanned by these elements $a$, is a Lie subalgebra of

$C^+(V)$, and $\tau' : \mathfrak{spin}(V) \to \mathfrak{so}(V)$ is an isomorphism of Lie algebras. The *spin group* $\mathrm{Spin}(V)$ of $(V, Q)$ is now defined as the connected Lie subgroup of $C^+(V)^\times$, with Lie algebra equal to $\mathfrak{spin}(V)$.

The tangent map at the identity of the homomorphism

$$\tau : a \mapsto \left(v \mapsto a \cdot v \cdot a^{-1}\right) \tag{4.2}$$

is equal to the isomorphism $\tau'$ from $\mathfrak{spin}(V)$ onto $\mathfrak{so}(V)$ defined in (4.1), so $\tau$ is a covering of Lie groups from $\mathrm{Spin}(V)$ onto $\mathrm{SO}(V)$. If $\xi, \eta \in V$, $Q(\xi, \xi) = Q(\eta, \eta) = 1$ and $Q(\xi, \eta) = 0$, then

$$(\xi \cdot \eta)^2 = \xi \cdot \eta \cdot \xi \cdot \eta = -\xi^2 \cdot \eta^2 = -1 \cdot -1 \cdot -1 = -1$$

Since

$$\xi \cdot \eta = -\eta \cdot \xi = \tfrac{1}{2}\,(\xi \cdot \eta - \eta \cdot \xi) \in \mathfrak{spin}(V),$$

we get the element

$$\begin{aligned} \exp_C\,(t\,\xi \cdot \eta) &= \sum_{k=0}^{\infty} \tfrac{t^k}{k!}\,(\xi \cdot \eta)^k \\ &= \cos t + (\sin t)\;\xi \cdot \eta \end{aligned} \tag{4.3}$$

in the group $\mathrm{Spin}(V)$. Taking $t = \pi$, we see that $-1 \in \mathrm{Spin}(V)$. Since $\tau(-1) = 1$, the covering is not an isomorphism. Since the universal covering of $\mathrm{SO}(V)$ is two-fold if $\dim(V) > 2$, we see that in this case $\mathrm{Spin}(V)$ is simply connected and isomorphic to the universal covering of $\mathrm{SO}(V)$.

The mapping $c : \mathrm{Spin}(V) \to \mathrm{End}(E)$ defines a representation of $\mathrm{Spin}(V)$ in $E$. Since $\mathrm{Spin}(V) \subset C^+(V)$, we see from (3.28) and (3.29) that $c\,(\mathrm{Spin}(V))$ leaves $E^+$ and $E^-$ invariant. It is one of the points in Brauer and Weyl [15], that the representations of $\mathrm{Spin}(V)$ in $E^+$ and $E^-$ are irreducible, and have highest weight equal to the fundamental ones, which are not equal to the highest weights of representations of $\mathrm{SO}(V)$.

## 4.2 The Spin-c Group

The unitary group $\mathrm{U}(V)$ of $V$ is defined as the group of the elements of $\mathrm{SO}(V)$, which commute with $J$. If $A \in \mathrm{U}(V)$, then its complex linear extension in $V \otimes \mathbf{C}$, also denoted by $A$, leaves the subspace $V^{(0,1)}$ of $V \otimes \mathbf{C}$ invariant. The restriction of $A$ to $V^{(0,1)}$ has a unique extension $\Lambda(A)$ to an automorphism of the exterior algebra $E = \Lambda V^{(0,1)}$, which actually coincides with the mapping $\Lambda(A)$ defined in (3.40). Using the isomorphism

$$\mathrm{c} : \mathrm{C}(V) \otimes \mathbf{C} \xrightarrow{\sim} \mathrm{End}(E), \tag{4.4}$$

we get an embedding $\mathrm{c}^{-1} \circ \Lambda$ of $\mathrm{U}(V)$ as a subgroup $U$ of $\mathrm{C}(V) \otimes \mathbf{C}$.

In order to identify the Lie algebra of $U$, we investigate, for each $\xi \in V$, cf. (3.20), (3.21):

$$\begin{aligned}
\mathrm{c}(\xi)\, \mathrm{c}\,(J\xi) &= [\mathrm{e}\,(\xi - iJ\xi) - \mathrm{i}\,(Q\xi)] \circ [\mathrm{e}\,(J\xi + i\xi) - \mathrm{i}\,(QJ\xi)] \\
&= [\mathrm{e}\,(\xi - iJ\xi) - \mathrm{i}\,(Q\xi)] \circ [i\ \mathrm{e}\,(\xi - iJ\xi) + i\ \mathrm{i}\,(Q\xi)] \\
&= i\ \mathrm{e}\,(\xi - iJ\xi) \circ \mathrm{i}\,(Q\xi) - i\ \mathrm{i}\,(Q\xi) \circ \mathrm{e}\,(\xi - iJ\xi) \\
&= 2i\ \mathrm{e}\,(\xi - i\,J\xi) \circ \mathrm{i}\,(Q\xi) - i\,Q\,(\xi,\, \xi)\,.
\end{aligned}$$

Here we have used that $\mathrm{i}(Jw) = -i\ \mathrm{i}(w)$ on antilinear forms in combination with $QJ = -QJ' = JQ$ (which follows from the antisymmetry of $J$ with respect to $Q$).

Now for any $\alpha \in V^{(0,1)}$ and $w \in W$, the operator $\mathrm{e}(\alpha) \circ \mathrm{i}(w)$ is a derivation of the exterior algebra $E$ of $V^{(0,1)}$, which on $V^{(0,1)}$ acts by sending $\omega$ to $\langle w,\, \omega\rangle\, \alpha$. So, if $a = \xi \cdot J\xi + i\,Q(\xi,\, \xi)$, then $\mathrm{c}(a)$ is equal to $\Lambda'(A)$, the derivation of $E$, induced by some linear transformation $A$ in $W$, which commutes with $J$. Since

$$\xi \cdot J\xi = -J\xi \cdot \xi = \tfrac{1}{2}\,(\xi \cdot J\xi - J\xi \cdot \xi)\,,$$

we see from (4.1) that

$$\tau'(a) : \zeta \mapsto 2\,[Q\,(\xi,\, \zeta)\ J\xi - Q\,(J\xi,\, \zeta)\ \xi]$$

belongs to the Lie algebra $\mathrm{u}(V)$ of $\mathrm{U}(V)$.

Running through the various identifications, we see that

$$\mathrm{c}(a) = \Lambda'\left(\tau'(a)\right). \tag{4.5}$$

We also get that $\tau'(a)$, considered as a complex linear transformation in $V$, has complex trace equal to $2i\, Q(\xi,\, \xi)$. Therefore

$$a - \tfrac{1}{2}\,\mathrm{trace}_{\mathbf{C}}\, \tau'(a) = a - i\, Q\,(\xi,\, \xi) = \xi \cdot J\xi \in \mathfrak{spin}(V). \tag{4.6}$$

It follows from (4.6), that thc Lie algebra of

$$U := \mathrm{c}^{-1} \circ \Lambda\left(\mathrm{U}(V)\right) \tag{4.7}$$

is contained in $\mathfrak{spin}(V) + \mathrm{u}(1)$, where $\mathrm{u}(1) = i\,\mathbf{R}$ is the Lie algebra of $\mathrm{U}(1)$, the unit circle in $\mathbf{C}$, viewed as a subset of $\mathrm{C}(V) \otimes \mathbf{C}$. So if we *define*

$$\mathrm{Spin}^{\mathrm{c}}(V) := \mathrm{Spin}(V) \cdot \mathrm{U}(1), \tag{4.8}$$

then we see that $U \subset \mathrm{Spin}^{\mathrm{c}}(V)$, and actually $\mathrm{Spin}^{\mathrm{c}}(V)$ is the group generated by $U$ and $\mathrm{Spin}(V)$. Note that $\mathrm{U}(1)$ commutes with everything, so $\mathrm{Spin}(V) \cap \mathrm{U}(1)$ belongs to the center of $\mathrm{Spin}(V)$, which is equal to $\{\pm 1\}$, because the center of $\mathrm{SO}(V)$ is trivial. It follows that the product map $p : (s,\, z) \mapsto s \cdot z$ induces an isomorphism

$$\mathrm{Spin}(V) \times \mathrm{U}(1)/\{\pm 1\} \xrightarrow{\sim} \mathrm{Spin}^{\mathrm{c}}(V), \tag{4.9}$$

where $-1$ denotes the element $(-1,\, -1)$ in $\mathrm{Spin}(V) \times \mathrm{U}(1)$. One sometimes sees the left hand side of (4.9) as the definition of $\mathrm{Spin}^{\mathrm{c}}(V)$, but we prefer to have $\mathrm{Spin}^{\mathrm{c}}(V)$ defined as a subgroup of $\left(\mathrm{C}^+(V) \otimes \mathbf{C}\right)^{\times}$, the group of invertible elements in $\mathrm{C}^+(V) \otimes \mathbf{C}$.

The injection of $\mathrm{U}(1)$ into $\mathrm{Spin}^{\mathrm{c}}(V)$, followed by the projection $\pi$ from $\mathrm{Spin}^{\mathrm{c}}(V)$ onto $\mathrm{Spin}^{\mathrm{c}}(V)/\mathrm{Spin}(V)$, has kernel equal to $\pm 1$. This means that there is a unique isomorphism

$$\iota : \mathrm{Spin}^{\mathrm{c}}(V)/\mathrm{Spin}(V) \xrightarrow{\sim} \mathrm{U}(1), \tag{4.10}$$

such that

$$\iota\,(z \cdot \mathrm{Spin}(V)) = z^2, \quad z \in \mathrm{U}(1). \tag{4.11}$$

We will also write $\iota(s) = \iota\,(s \cdot \mathrm{Spin}(V))$, if $s \in \mathrm{Spin}^{\mathrm{c}}(V)$. Finally, the homomorphism (4.2) extends to the homomorphism

$$\tau : a \mapsto \left(v \mapsto a \cdot v \cdot a^{-1}\right) : \mathrm{Spin}^{\mathrm{c}}(V) \to \mathrm{SO}(V), \tag{4.12}$$

which has kernel equal to $\mathrm{U}(1)$, the unit circle in $\mathbf{C} \subset \mathrm{C}(V) \otimes \mathbf{C}$.

Integrating (4.5), we get

$$\mathrm{c}(u) = \Lambda \circ \tau(u), \quad u \in U. \tag{4.13}$$

In turn, this implies that $\tau|_U$ is an isomorphism from $U$ onto $\mathrm{U}(V)$, with inverse equal to $\mathrm{c}^{-1} \circ \Lambda$. The equation (4.6) shows that

$$\{z \cdot u \mid u \in U,\ z \in \mathbf{C},\ z^2 \det_{\mathbf{C}} \tau(u) = 1\} \subset \mathrm{Spin}(V) \tag{4.14}$$

is equal to the preimage in $\mathrm{Spin}(V)$ of $\mathrm{U}(V)$ under the double covering $\tau : \mathrm{Spin}(V) \to \mathrm{SO}(V)$.

## 4.3 Proof of a Formula for the Supertrace

As an application of the previous section, we give a proof of the formula (3.33), which expreses the supertrace in terms of the volume part. Let $\epsilon_j$ be a unitary frame in $V$; consider the element

$$a = \sum_{j=1}^{n} \epsilon_j \cdot J\epsilon_j + i\,n \in \mathfrak{u}.$$

Then (4.6) shows that $\mathrm{c}(a)$ is equal to the derivation of $E$, which acts as multiplication by $2i$ on $E^1$. It follows that

$$\mathrm{str}_{\mathbf{C}}\,\mathrm{c}\,(\exp_{\mathrm{C}}(t\,a)) = \sum_{q=0}^{n}(-1)^q \binom{n}{q} \mathrm{e}^{2iqt} = \left(1 - \mathrm{e}^{2it}\right)^n.$$

On the other hand, we read from (4.3) that

$$\exp_{\mathbb{C}}(t\,a) = \mathrm{e}^{itn} \prod_{j=1}^{n} (\cos t + \sin t\, \epsilon_j \cdot J\epsilon_j)\,,$$

so

$$\operatorname{vol}\left(\sigma_\Lambda\left(\exp_{\mathbb{C}}(t\,a)\right)\right) = \mathrm{e}^{itn}\,(\sin t)^n = (2i)^{-n}\left(\mathrm{e}^{2it} - 1\right)^n.$$

Comparing the two expressions now yields (3.33).

# Chapter 5

# The Spin-c Dirac Operator

In this chapter we start by viewing $E$ as a principal bundle for the group $\mathrm{Spin}^{\mathrm{c}}(2n)$, which contains the unitary group. The fact that $\mathrm{Spin}^{\mathrm{c}}(2n)$ also contains the spin group $\mathrm{Spin}(2n)$, which is a twofold cover of $\mathrm{SO}(2n)$, allows us to introduce a connection in this principal bundle which has the desired compatibility with the Levi-Civita connection. Using this connection in $E$, we will give the definition of the spin-c Dirac operator $D$ in (5.14). In Lemma 5.5 it is established that $D$ is selfadjoint and has the same principal symbol as the Dolbeault-Dirac operator.

## 5.1 The Spin-c Frame Bundle and Connections

The principal $\mathrm{Spin}^{\mathrm{c}}(2n)$-bundle over $M$, which will replace the unitary frame bundle $\mathrm{UF}\,M$ in (3.42), is the *spin-c frame bundle*

$$\mathrm{Spin}^{\mathrm{c}}\,\mathrm{F}\,M = \mathrm{UF}\,M \times_{\mathrm{U}(n)} \mathrm{Spin}^{\mathrm{c}}(2n), \tag{5.1}$$

where the action of $\mathrm{U}(n)$ on $\mathrm{UF}\,M \times \mathrm{Spin}^{\mathrm{c}}(2n)$ is given by

$$(e,\, r) \mapsto \left(e \circ A^{-1},\, \mathrm{c}^{-1} \circ \Lambda(A) \cdot r\right), \quad A \in \mathrm{U}(n). \tag{5.2}$$

Admittedly, the name is not ideal, because the elements in the fiber, the "spin-c frames", do not have a straightforward interpretation as bases in some vector space. Since the action of $\mathrm{U}(n)$ on $\mathrm{UF}\, M$ is free, the free action of $\mathrm{Spin}^{\mathrm{c}}(2n)$ on itself by right multiplication passes to a free action of $\mathrm{Spin}^{\mathrm{c}}(2n)$ on $\mathrm{Spin}^{\mathrm{c}}\,\mathrm{F}\, M$, exhibiting the latter as a principal $\mathrm{Spin}^{\mathrm{c}}(2n)$-bundle over $M$.

Since the Levi-Civita connection of the oriented orthonormal frame bundle $\mathrm{SOF}\, M$ in general does not restrict to the subbundle $\mathrm{UF}\, M$, we will not use the definition (5.1) in order to define the connection in $\mathrm{Spin}^{\mathrm{c}}\,\mathrm{F}\, M$. Instead, we consider the mapping

$$\alpha : (e,\, r) \mapsto (f(e),\, \tau(r)) : \mathrm{UF}\, M \times \mathrm{Spin}^{\mathrm{c}}(2n) \to \mathrm{SOF}\, M \times \mathrm{SO}(2n) \quad (5.3)$$

with $f(e)$ defined as in (3.44), and the mapping

$$\beta : (e,\, r) \mapsto (e,\, \iota(r)) : \mathrm{UF}\, M \times \mathrm{Spin}^{\mathrm{c}}(2n) \to \mathrm{UF}\, M \times \mathrm{U}(1). \quad (5.4)$$

Using (4.13), we see that $\alpha$ maps $\mathrm{U}(n)$-orbits into $\mathrm{SO}(2n)$-orbits for the action

$$A \mapsto \Big( (f,\, B) \mapsto \big( f \circ A^{-1},\, A \circ B \big) \Big).$$

The quotient for the latter action is isomorphic to $\mathrm{SOF}\, M$, so $\alpha$ induces a bundle mapping

$$\alpha : \mathrm{Spin}^{\mathrm{c}}\,\mathrm{F}\, M \to \mathrm{SOF}\, M.$$

On the other hand, combining (4.11) and (4.14), we see that $\beta$ maps $\mathrm{U}(n)$-orbits in $\mathrm{UF}\, M \times \mathrm{Spin}^{\mathrm{c}}(2n)$ into $\mathrm{U}(n)$-orbits in $\mathrm{UF}\, M \times \mathrm{U}(1)$ for the action

$$A \mapsto \Big( (e,\, z) \mapsto \big( e \circ A^{-1},\, z \, \det\nolimits_{\mathbf{C}} A \big) \Big). \quad (5.5)$$

The quotient of $\mathrm{UF}\, M \times \mathrm{U}(1)$ by the $\mathrm{U}(n)$-action is a principal $\mathrm{U}(1)$-bundle. It can be identified with the unit circle bundle $\mathrm{U}\,(K^*)$ of the dual $K^*$ of the *canonical line bundle* $K = \mathrm{T}^*\, M^{(n,\,0)}$ of $(M,\, J)$. Indeed, an element $\omega \in \mathrm{T}_x^*\, M^{(n,\,0)}$ can be identified with the mapping

$$e \mapsto \omega\,(e_1,\, \ldots,\, e_n) : \mathrm{UF}_x\, M \to \mathbf{C},$$

which satisfies

$$\omega\left(e \circ A^{-1}\right) = \det_{\mathbf{C}}\left(e \circ A^{-1} \circ e^{-1}\right)\, \omega(e) = \left(\det_{\mathbf{C}} A\right)^{-1}\, \omega(e).$$

The mappings $\alpha$ and $\beta$ together define a bundle map

$$\gamma : \mathrm{Spin^c\,F}\,M \to \Delta^* \left(\mathrm{SOF}\,M \times \mathrm{U}(K^*)\right), \tag{5.6}$$

where $\Delta : x \mapsto (x,\, x)$ denotes the embedding of $M$ onto the diagonal in $M \times M$. If $r \in \mathrm{Spin^c}(2n)$, $\tau(r) = 1$ and $\iota(r) = 1$, then $r = z \in \mathrm{U}(1)$ and (4.11) shows that $r = \pm 1$. Using this, one obtains that (5.6) is a double covering.

Now choose a $\mathrm{U}(1)$-invariant connection in the principal $\mathrm{U}(1)$-bundle $\mathrm{U}(K^*)$. (This is the same as a linear connection in the line bundle $K^*$ for which the Hermitian structure in $K^*$ is covariantly constant.) Together with the Levi-Civita connection in $\mathrm{SOF}\,M$, we get a connection in the product bundle, which then is pulled back by $\Delta$ and $\gamma$ to a connection in the spin-c frame bundle. That is, the horizontal space in the tangent space at $p$ of $\mathrm{Spin^c\,F}\,M$ consists of the vectors $v$, such that $\mathrm{T}_p\,\alpha(v)$ is horizontal in the tangent space at $\alpha(p)$ of $\mathrm{SOF}\,M$, *and* $\mathrm{T}_p\,\beta(v)$ is horizontal in $\mathrm{U}\left(K^*\right)$. The first condition determines $v$ up to the addition of a scalar in the fiber direction, whose freedom then is eliminated by the second condition.

Since $\alpha$ and $\beta$ are right $\mathrm{Spin^c}(2n)$-equivariant, the connection is right $\mathrm{Spin^c}(2n)$-invariant. Here the action on $\mathrm{SOF}\,M$ is the right action of $\mathrm{SO}(2n)$, via the homomorphism $\tau$, defined in (4.12). And the action on $\mathrm{U}(K^*)$ is the right action of $\mathrm{U}(1)$, via the homomorphism $\iota$, defined in (4.10).

The mapping $e \mapsto (e,\, 1)$ defines an embedding of $\mathrm{UF}\,M$ into $\mathrm{Spin^c\,F}\,M$. We note in passing that if $(M,\, J,\, h)$ is a Kähler manifold, then the fact that the Levi-Civita connection in $\mathrm{SOF}\,M$ is tangent to the subbundle $\mathrm{UF}\,M$ implies that the connection in $\mathrm{Spin^c\,F}\,M$, which we just defined, is tangent to the subbundle $\mathrm{UF}\,M$ of $\mathrm{Spin^c\,F}\,M$.

**Lemma 5.1** *The embedding*

$$\mathrm{UF}M \times \mathrm{E}(n) \to \mathrm{Spin^cF}M \times \mathrm{E}(n)$$

*leads to an isomorphism*

$$E = \mathrm{UF}M \times_{\mathrm{U}(n)} \mathrm{E}(n) \xrightarrow{\sim} \mathrm{Spin^c F}M \times_{\mathrm{Spin^c}(2n)} \mathrm{E}(n), \qquad (5.7)$$

*if, in* $\mathrm{UF}M \times \mathrm{E}(n)$*, the* $\mathrm{U}(n)$*-action is given by (3.43) and in* $\mathrm{Spin^cF}M \times \mathrm{E}(n)$ *the action of* $s \in \mathrm{Spin^c}(2n)$ *is given by*

$$(g,\, \omega) \mapsto (s \cdot g,\, \mathrm{c}(s)(\omega))\,.$$

**Proof** The $\mathrm{Spin^c}(2n)$-orbits in $\mathrm{Spin^c\,F}\,M \times \mathrm{E}(n)$ can be identified with the $\mathrm{U}(n) \times \mathrm{Spin^c}(2n)$-orbits in

$$\mathrm{UF}\,M \times \mathrm{Spin^c}(2n) \times \mathrm{E}(n),$$

where the action of

$$(A,\, s) \in \mathrm{U}(n) \times \mathrm{Spin^c}(2n)$$

is given by

$$(e,\, r,\, \omega) \mapsto \left(e \circ A^{-1},\, \mathrm{c}^{-1} \circ \Lambda(A) \cdot r \cdot s^{-1},\, \mathrm{c}(s)(\omega)\right).$$

If $r = 1$ and $\mathrm{c}^{-1} \circ \Lambda(A) \cdot 1 \cdot s^{-1} = 1$, then $\Lambda(A) = \mathrm{c}(s)$. This proves (5.7). □

This means that the embedding of $\mathrm{UF}\,M$ into $\mathrm{Spin^c\,F}\,M$ exhibits $E$ as a vector bundle which is associated to the principal $\mathrm{Spin^c}(2n)$-bundle $\mathrm{Spin^c\,F}\,M$ by means of the representation

$$\mathrm{c} : \mathrm{Spin^c}(2n) \to \mathrm{End}\,(\mathrm{E}(n))\,.$$

In the same way as in (3.45), the $\mathrm{Spin^c}(2n)$-invariant connection in $\mathrm{Spin^c\,F}\,M$ leads to a covariant differentiation of sections of $E$, which we again denote by $\nabla$. If $(M,\, J,\, h)$ is a Kähler manifold, then $\nabla$ is equal to the covariant differentiation defined in (3.45).

**Lemma 5.2** *The Clifford algebra bundle* $\mathrm{C}\,(\mathrm{T}^* M)$ *is isomorphic to the associated* $\mathrm{Spin}^{\mathrm{c}}(2n)$*-bundle*

$$\mathrm{Spin}^{\mathrm{c}}\mathrm{F}M \times_{\mathrm{Spin}^{\mathrm{c}}(2n)} \mathrm{C}(2n), \tag{5.8}$$

*where the action of* $s \in \mathrm{Spin}^{\mathrm{c}}(2n)$ *on* $\mathrm{Spin}^{\mathrm{c}}\mathrm{F}M \times \mathrm{C}(2n)$ *is given by*

$$(p,\, a) \mapsto \left(s\,p,\; s \cdot a \cdot s^{-1}\right) = (s\,p,\; \tau(s)(a))\,. \tag{5.9}$$

**Proof** Using the definition (5.1) of the principal $\mathrm{Spin}^{\mathrm{c}}(2n)$-bundle $\mathrm{Spin}^{\mathrm{c}}\,\mathrm{F}\,M$, we see that (5.8) is equal to the space of orbits in

$$\mathrm{UF}\,M \times \mathrm{Spin}^{\mathrm{c}}(2n) \times \mathrm{C}(2n)$$

for the action of $A \in \mathrm{U}(n)$, $s \in \mathrm{Spin}^{\mathrm{c}}(2n)$, given by

$$(e,\, r,\, a) \mapsto \left(e \circ A^{-1},\; \mathrm{c}^{-1} \circ \Lambda(A) \cdot r \cdot s^{-1},\; \tau(s)(a)\right).$$

If $s = \mathrm{c}^{-1} \circ \Lambda(A)$, then $s \in U$, so (4.13) yields that

$$\Lambda \circ \tau(s) = \mathrm{c}(s) = \Lambda(A),$$

which implies that $\tau(s) = A$. So the embedding $(e,\, a) \to (e,\, 1,\, a)$ induces an isomorphism

$$\mathrm{C}\,(\mathrm{T}^*\, M) = \mathrm{UF}\,M \times_{\mathrm{U}(n)} \mathrm{C}(2n) \xrightarrow{\sim} \mathrm{Spin}^{\mathrm{c}}\,\mathrm{F}\,M \times_{\mathrm{Spin}_{\mathbf{c}}(2n)} \mathrm{C}(2n). \tag{5.10}$$

This is the isomorphism meant in the lemma. □

Again, as in (3.45), the $\mathrm{Spin}^{\mathrm{c}}(2n)$-invariant connection in the spin-c frame bundle leads to a covariant differentiation $\nabla$ of sections of $\mathrm{C}\,(\mathrm{T}^*\, M)$. Fortunately, we have:

**Lemma 5.3** *The covariant differentiation in (5.8) is equal to the covariant differentiation in (3.37). In each case the covariant differentiation is defined by (3.38). The horizontal lifts of vector fields in $M$ are with regard to the previously defined connection in* $\mathrm{Spin}^{\mathrm{c}}\mathrm{F}M$ *and the Levi-Civita connection in* $\mathrm{SOF}M$*, respectively.*

**Proof** Sections of $\mathrm{C}\,(\mathrm{T}^*\,M)$, viewed as a principal $\mathrm{Spin}^{\mathrm{c}}(2n)$-bundle, are mappings $a$ from $\mathrm{UF}\,M \times \mathrm{Spin}^{\mathrm{c}}(2n)$ to $\mathrm{C}(2n)$ such that, for each $e \in \mathrm{UF}\,M$, $r,\ s \in \mathrm{Spin}^{\mathrm{c}}(2n)$, and $A \in \mathrm{U}(n)$:

$$a\left(e \circ A^{-1},\ \mathrm{c}^{-1} \circ \Lambda(A) \cdot r \cdot s^{-1}\right) = \tau(s)\ a\,(e,\ r)\,.$$

This implies in particular that $a\,(e,\ -r) = a\,(e,\ r)$. Therefore, using (5.6), $a$ can be identified with a mapping from $\mathrm{SOF}_x\,M \times \mathrm{UF}_x\,M \times \mathrm{U}(1)$ to $\mathrm{C}(2n)$, such that, for each $f \in \mathrm{SOF}_x\,M$, $e \in \mathrm{UF}_x\,M$, $z \in \mathrm{U}(1)$, and $A \in \mathrm{U}(n)$, $s \in \mathrm{Spin}^{\mathrm{c}}(2n)$:

$$a\left(f \circ \tau(s)^{-1},\ e \circ A^{-1},\ \frac{z}{\iota(s)}\ \det\nolimits_{\mathbf{C}} A\right) = \tau(s)\ a\,(f,\ e,\ z)\,.$$

From the fact that

$$\tau\,(y \cdot s) = \tau\,(s)$$

for any $y \in \mathbf{C}$, we see that $a\,(f,\ e,\ z)$ does not depend on $z$. It follows in turn that it does not depend on $e$ either. So the covariant derivative of $a$ is equal to the horizontal derivative of $a$, viewed as a mapping from $\mathrm{SOF}\,M$ to $\mathrm{C}(2n)$, which satisfies, for each $f \in \mathrm{SOF}\,M$ and $A \in \mathrm{SO}(2n)$:

$$a\left(f \circ A^{-1}\right) = A\,(a(f))\,. \tag{5.11}$$

That is, $a$ is viewed as a section of (3.37). □

Using the Leibniz rule (3.46), we arrive at the following conclusion.

**Lemma 5.4** *Choose any* $\mathrm{U}(1)$*-invariant connection in* $\mathrm{U}(K^*)$*, the unit circle bundle in the dual* $K^*$ *of the canonical line bundle* $K$ *of* $(M, J)$*. Then the covariant differentiation* $\nabla$ *on* $E$*, defined by the Levi-Civita connection in* $\mathrm{SOF}M$ *and the connection in* $\mathrm{U}(K^*)$*, is a Clifford connection.*

If $L$ is a complex vector bundle over $M$, then the $L$-valued $(0, \cdot)$-forms are the sections of $E \otimes L$. We assume that a connection in $L$ is chosen which leaves the Hermitian structure $h^L$ in $L$ invariant, and denote the corresponding covariant differentiation of sections of $L$ by $\nabla^L$. Then the covariant differentiation $\nabla$ of sections of $E \otimes L$ (which depends on the choice of $\nabla^L$, although this will be suppressed in the notation) is defined by the Leibniz rule

$$\nabla_v \left( \sum_{j,k} \omega_{jk} e_j \otimes l_k \right) = \sum_{j,k} (v\omega_{jk})\, e_j \otimes l_k + \omega_{jk} v_{\mathrm{hor}} e_j \otimes l_k + \omega_{jk} e_j \otimes \nabla^L l_k. \tag{5.12}$$

Here $v$ is a vector field in $M$, the $e_j$ and $l_k$ form an arbitrary local frame in $E$ and $L$, respectively, and the complex-valued functions $\omega_{jk}$ are the coefficients of $\omega$ with respect to the local frame $e_j \otimes l_k$ in $E \otimes L$. A straightforward calculation shows that the right hand side in (5.12) is independent of the choice of local frames, so the local definitons (5.12) piece together to a global differential operator on $M$. It is also clear that $\nabla$ is determined by the conditions that it is a covariant differentiation and that

$$\nabla_v (\omega \otimes \lambda) = v_{\mathrm{hor}} \omega \otimes \lambda + \omega \otimes \nabla^L \lambda, \tag{5.13}$$

for every local section $\omega$ and $\lambda$ of $E$ and $L$, respectively.

## 5.2 Definition of the Spin-c Dirac Operator

At long last, we are ready for the definition of the spin-c Dirac operator, for which we want to have the local heat kernel formula.

**Definition** Given connections in the dual $K^*$ of the canonical line bundle $K = \mathrm{T}^* M^{(n,0)}$, and in the complex vector bundle $L$, the spin-c Dirac operator $D$, acting on sections of $E \otimes L$, is given by

$$D\,\omega = \sum_{j=1}^{2n} \mathrm{c}\,(\phi_j)\left(\nabla_{f_j}\omega\right). \tag{5.14}$$

Here $(f_j)_{j=1}^{2n}$ is an arbitrary (not necessarily oriented or orthonormal) local frame in $\mathrm{T}\,M$ and $(\phi_j)_{j=1}^{2n}$ is the corresponding dual frame. That is

$$\langle f_j,\, \phi_k\rangle = \delta_{jk}.$$

Since (5.14) does not involve differentiations of the frames, it is independent of the choice of the local frames, so the local definitions (5.14) lead to a globally defined operator $D$ in $\Gamma\,(M,\, E \otimes L)$.

**Remark**

A straightforward generalization can be given as follows. A *spin-c structure* on $M$ is a principal $\mathrm{Spin}^{\mathrm{c}}(2n)$-bundle $P$ over $M$, together with a complex line bundle $\tilde{K}$ with Hermitian structure and connection over $M$, and a twofold covering from $P$ onto $Q = \Delta^*\left(\mathrm{SOF}\,M \times \mathrm{U}(\tilde{K})\right)$, which intertwines the action of $r \in \mathrm{Spin}^{\mathrm{c}}(2n)$ on $P$ with the action of $(\tau(r),\, \iota(r))$ on $Q$. All the previous definitions can now be repeated, in which $K^*$ is replaced by $\tilde{K}$. This leads to a Dirac operator, acting on sections of $E \otimes L$, in which now

$$E = P \times_{\mathrm{Spin}^{\mathrm{c}}} \mathrm{E}(n).$$

The reader may verify that most of the results in the sequel carry over to this more general operator $D$, just by replacing $K^*$ by $\tilde{K}$ at every occasion. However, the relation with the almost complex structure is lost.

For a compact oriented manifold $M$ of any dimension, a spin-c structure exists on $M$ if and only if the Stiefel-Whitney class $w_2 \in \mathrm{H}^2(M,\, \mathbf{Z}/2\mathbf{Z})$ is equal to the reduction modulo two of some $c \in \mathrm{H}^2(M,\, \mathbf{Z})$. Hirzebruch and Hopf [40, 4.1,iv)] proved that every four-dimensional compact oriented manifold $M$ satisfies this condition; it therefore can be provided with a spin-c structure. On the other hand, it is a result of Wu Wen-Tsun [76, Th. 10, p.74] that a four-dimensional compact oriented manifold $M$ can be provided with an almost complex structure, if and only if there exists a class $c \in \mathrm{H}^2(M, \mathbf{Z})$ such that $w_2 = c \bmod 2$ and $c^2 = 2e + p_1$. Here $e \in \mathrm{H}^4(M, \mathbf{Z})$ and $p_1 \in \mathrm{H}^4(M, \mathbf{Z})$ are the Euler class and the first Pontryagin class of $M$, respectively. See also [40, Th. 4.6 and 4.1,ii)].

For some more details on spin-c structures, cf. Lawson and Michelsohn [51, Appendix D]

**Lemma 5.5** *The spin-c Dirac operator D is selfadjoint. Its principal symbol is given by*

$$\sigma_D(\xi) = i\,\mathrm{c}(\xi), \tag{5.15}$$

*which in turn is equal to the principal symbol of the Dolbeault-Dirac operator* $2\left(\bar{\partial} + \bar{\partial}^*\right)$.

**Proof** Let $\omega$ and $\nu$ be sections, one of which has compact support contained in an open subset of $M$ in which we have a local frame $f_j$. Then we get, with the notation (5.14):

$$\begin{aligned}
(\omega,\, D^*\nu) = (D\omega,\, \nu) &= \sum_j \int_M h\left(\mathrm{c}\left(\phi_j\right)\, \nabla_{f_j}\omega,\, \nu\right)\, \mathrm{d}x \\
&= -\sum_j \int_M h\left(\nabla_{f_j}\omega,\, \mathrm{c}\left(\phi_j\right)\, \nu\right)\, \mathrm{d}x \\
&= -\sum_j \int_M f_j\left(h\left(\omega,\, \mathrm{c}\left(\phi_j\right)\nu\right)\right) - h\left(\omega,\, \nabla_{f_j}\mathrm{c}\left(\phi_j\right)\, \nu\right)\, \mathrm{d}x.
\end{aligned}$$

For any vector field $v$, the *divergence* of $v$ with respect to the volume form $\mathrm{d}x$ is the function $\operatorname{div} v$, defined by the relation

$$\mathcal{L}(v)\,(\mathrm{d}x) = (\operatorname{div} v)\, \mathrm{d}x.$$

Here $\mathcal{L}(v)$ denotes the Lie derivative with respect to $v$. With this notation, we get for any function $g$ that

$$\int_M f_j\, g \;\mathrm{d}x = -\int_M g \;\mathrm{div}\, f_j \;\mathrm{d}x.$$

Using also (3.39), we arrive at

$$D^* = D + \sum_j \mathrm{div}\, f_j \;\mathrm{c}\,(\phi_j) + \mathrm{c}\left(\nabla_{f_j}\phi_j\right)$$

for the adjoint of $D$.

If the $f_j$ form an oriented orthonormal frame, then

$$\mathrm{d}x = \phi_1 \wedge \phi_2 \wedge \ldots \wedge \phi_{2n},$$

so

$$\mathcal{L}\,(f_j)\,\mathrm{d}x = \sum_k \phi_1 \wedge \ldots \wedge \mathcal{L}\,(f_j)\,\phi_k \wedge \ldots \wedge \phi_{2n}.$$

On the other hand,

$$\langle f_l,\, \mathcal{L}\,(f_j)\;\phi_k\rangle = -\langle \mathcal{L}\,(f_j)\; f_l,\, \phi_k\rangle,$$

combined with the assumption that the connection is torsion-free, shows that

$$\begin{aligned}\mathcal{L}\,(f_j)\;\phi_k &= -\sum_l \langle \mathcal{L}\,(f_j)\, f_l,\, \phi_k\rangle\, \phi_l \\ &= -\sum_l \langle \nabla_{f_j} f_l - \nabla_{f_l} f_j,\, \phi_k\rangle\, \phi_l.\end{aligned}$$

For any $x \in M$, there exists a local oriented orthonormal frame $f_j$, which is horizontal at $x$. It follows that $\left(\nabla_{f_j}\phi_j\right)(x) = 0$ and $(\mathrm{div}\, f_j)\,(x) = 0$, so $(D^*\omega)\,(x) = (D\omega)\,(x)$.

The formula (5.15) for the principal symbol follows immediately from the definition, and the last statement follows in view of (2.31), (2.34), and (3.24), with $Q$ replaced by $\beta^{-1}$. □

As for the Dolbeault-Dirac operator, we have (2.37) and (2.38), that is, $D$ switches $\Gamma(M, E^+ \otimes L)$ and $\Gamma(M, E^- \otimes L)$. And $D^-$ is equal to the adjoint of $D^+$.

If $M$ is compact, then the ellipticity of $D$ implies that ker $D^+$ and ker $D^-$ are finite-dimensional, and we have the index and virtual character of $D^+$, defined as in (2.39) and (2.40), respectively.

Now suppose for the moment that $M$ is a compact complex analytic manifold. Since the spin-c Dirac operator has the same principal symbol as the Dolbeault-Dirac operator, and the numbers (2.39) and (2.40) only depend on the (homotopy class of the) principal symbol of the operator $D^+$, it follows that these numbers for the spin-c Dirac operator are equal to the Riemann-Roch number and the holomorphic Lefschetz number, defined in (2.20) and (2.21), respectively.

# Chapter 6

# Its Square

The main goal of this chapter is Theorem 6.1, which says that the square of the spin-c Dirac operator $D$ is equal to the Laplace operator plus a zero order term, given by curvature expressions. The contribution from the curvature of $L$ will be responsible for the Chern characters $\mathrm{ch}\,(L_j)$ in Proposition 13.2. On the other hand, the term with one half of the curvature of $K^*$ leads, by combining the corresponding factors in (11.17) and (12.12) with the real determinants, to the complex determinants in Proposition 13.1 and Proposition 13.2, respectively. Theorem 6.1 is followed by a comparison of the spin-c Dirac operator with the spinor Dirac operator, which exists if $M$ is provided with a spin structure. We conclude this chapter with the description, in Proposition 6.1, of what happens with the formula for $D^2$ in the Kähler case when $D$ is equal to the Dolbeault-Dirac operator.

We begin by repeating some generalities about the Laplacian and the curvature of a connection. If $\nabla$ is any covariant differentiation in a vector bundle over a Riemannian manifold $(M,\ \beta)$, then the corresponding *Laplace operator* $\Delta$ is defined by

$$\Delta = -\sum_{j,k} \beta^{-1}(\phi_j, \phi_k)\left(\nabla_{f_j} \circ \nabla_{f_k} - \nabla_{\nabla^{\beta}_{f_j} f_k}\right). \tag{6.1}$$

Here $f_j$ is any local frame in $\mathrm{T}\,M$ and $\phi_j$ is the corresponding dual frame. (6.1) does not depend on the choice of the local frame, and therefore leads to a well-defined second order differential operator $\Delta$ acting on sections of the vector bundle over $M$. This is based on the observation that for any vector fields $v$ and $w$ in $M$ and any real-valued function $g$ in $M$, one has that

$$\nabla_v \circ \nabla_{g\,w} - \nabla_{\nabla^{\beta}_v (g\,w)} = g\left(\nabla_v \circ \nabla_w - \nabla_{\nabla^{\beta}_v w}\right).$$

The principal symbol of the Laplace operator $\Delta$ is given by

$$\sigma_{\Delta}(\xi) = \beta^{-1}(\xi, \xi). \tag{6.2}$$

And in the same way as in the proof of Lemma 5.5, one can prove that

$$\int_M h(\Delta\omega, \nu)\, \mathrm{d}x = \int_M \sum_j h\left(\nabla_{f_j}\omega, \nabla_{f_j}\nu\right) \mathrm{d}x = \int_M h(\omega, \Delta\nu)\, \mathrm{d}x, \tag{6.3}$$

for any smooth sections $\omega$ and $\nu$, one of which with compact support. In the middle expression the integrand is independent of the choice of local orthonormal frames $f_j$; it can be viewed as the inner product of the linear mappings $\nabla\omega$ and $\nabla\nu$. For $\omega = \nu$ the middle expression therefore is a Dirichlet integral, the square of the $\mathrm{L}^2$-norm of $\nabla\omega$. It can be used to give an alternative definition of $\Delta$.

Another important operator attached to a covariant differentiation $\nabla$ is its *curvature* $R = R^{\nabla}$, defined by

$$R(v, w) = \nabla_v \circ \nabla_w - \nabla_w \circ \nabla_v - \nabla_{[v,w]}. \tag{6.4}$$

Here $v$ and $w$ are vector fields in $M$ and $[v, w] = \mathcal{L}(v)\, w$ is the Lie bracket. The point is that, for any section $\omega$ and any point $x \in M$, the value of

$R(v, w)\,\omega$ at $x$ only depends on $v_x$, $w_x$, and $\omega_x$, so $R$ is of zero order as a differential operator in all its variables.

We will use the symbol $\nabla^\beta$ for the covariant differentiation of vector fields defined by the Levi-Civita connection of $(M, \beta)$. And we will use the symbol $R^\beta$ for the corresponding curvature, which at each $x \in M$ is a bilinear antisymmetric mapping from $\mathrm{T}_x M \times \mathrm{T}_x M$ to $\mathfrak{so}(\mathrm{T}_x M)$. If $f_j$ is a local orthonormal frame in $\mathrm{T} M$, then the coefficients of $R^\beta$ are denoted by

$$R^\beta_{mljk} = \beta\left(R^\beta(f_j, f_k)\, f_l, f_m\right). \tag{6.5}$$

Finally the *scalar curvature* $r^\beta$ of the Riemannian structure $\beta$ is the real-valued function on $M$, defined by

$$r^\beta = \sum_{j,k} R^\beta_{jkjk}. \tag{6.6}$$

It is easy to check that (6.6) does not depend on the choice of the local orthonormal frame $f_j$.

In the proof of Theorem 6.1, we need the relation which exists between the curvature form of a principal fiber bundle and the curvature operator of a vector bundle which is associated to it. In order to be able to write this down, we shortly recall the definition of the curvature form of a general principal fiber bundle.

If $G$ is a Lie group and $P$ is a principal $G$-bundle over $M$, then a *connection form* in $P$ is defined as a $G$-equivariant $\mathfrak{g}$-valued one-form $\theta$ in $P$, such that $\mathrm{i}\,(X_P)\,\theta = X$ for each $X \in \mathfrak{g}$, the Lie algebra of $G$. The equivariance is with respect to the adjoint action of $G$ in $\mathfrak{g}$. The $\ker\theta_p$, $p \in P$, define the horizontal spaces for a $G$-invariant connection in $P$; each $G$-invariant connection in $P$ arises in this way, and for a unique connection form $\theta$. The $G$-equivariant $\mathfrak{g}$-valued two-form $\Omega$ in $P$, defined by

$$\Omega(v, w) = (\mathrm{d}\theta)(v, w) - [\theta(v), \theta(w)] \tag{6.7}$$

for any vector fields $v$, $w$ in $P$, is called the *curvature form* of the connection form $\theta$. The curvature form $\Omega$ is *horizontal*, in the sense that $\mathrm{i}\,(X_P)\,\Omega = 0$, for any $X \in \mathfrak{g}$. Furthermore, if $v$ and $w$ are horizontal vector fields in $M$, then we see from (6.7) and (2.16) that

$$\Omega(v, w) = -\theta([v, w]).$$

If $\rho : G \to \mathrm{End}(V)$ is a representation of $G$ in the finite-dimensional complex vector space $V$, then we have the action

$$(p, v) \mapsto (g \cdot p, \rho(g) v)$$

of $g \in G$ in $P \times V$. The quotient $Q = P \times_G V$ is an associated vector bundle over $M$. For the sections of $Q$ we can define the covariant differentiation $\nabla$ as in (3.45) (with $E$ replaced by $Q$). The point of the previous paragraph is that its curvature operator $R^\nabla$ is given by

$$\begin{aligned} R^\nabla(v, w) &= [v_{\mathrm{hor}}, w_{\mathrm{hor}}] - [v, w]_{\mathrm{hor}} \\ &= -\Omega(v_{\mathrm{hor}}, w_{\mathrm{hor}})_P . \end{aligned} \tag{6.8}$$

Here $v$ and $w$ are vector fields in $M$, with horizontal lifts $v_{\mathrm{hor}}$ and $w_{\mathrm{hor}}$, respectively. In turn, because a section $\omega$ of $Q$ is identified with a mapping $\omega : P \to V$ such that

$$\omega(g \cdot p) = \rho(g)\,\omega(p)$$

for each $p \in P$ and $g \in G$, we see that

$$(X_P \omega)(p) = \rho'(X)\,\omega(p)$$

for each $p \in P$ and $X \in \mathfrak{g}$. Therefore we arrive at the following conclusion:

$$R^\nabla(v, w) = -\rho'\,\Omega(v_{\mathrm{hor}}, w_{\mathrm{hor}}). \tag{6.9}$$

In the literature, it also very customary to write the action of $g \in G$ on $P$ as a "right action" $p \mapsto g \cdot p$, instead of $p \mapsto g \cdot p$ or $p \mapsto p \cdot g^{-1}$, as we consistently have done. Such a right action defines an anti-homorphism from the group $G$ to the group of diffeomorphisms of $P$ instead of the homomorphism which we get. On the other hand, it has the advantage of leading to the formula (6.9) *without the minus sign*. This is one of the numerous sources of sign confusion in differential geometry when dealing with antisymmetric forms.

Returning to the Laplace operator $\Delta$, we see that its principal symbol $\sigma_\Delta$ coincides with the principal symbol

$$\sigma_{D^2}(\xi) = \sigma_D(\xi)^2$$

of $D^2$, cf. (5.15) and (3.25). This shows that $D^2 - \Delta$ is a differential operator of order at most equal to one. The careful choice of the covariant differentiation in $E$ now yields that $D^2 - \Delta$ actually is of zero order, and can be expressed in terms of the scalar curvature of $(M, \beta)$, and terms in which the curvatures $R^{K^*}$ and $R^L$ of the covariant differentiations $\nabla^{K^*}$ and $\nabla^L$ of the line bundle $K^*$ and the vector bundle $L$, respectively, are combined with the Clifford multiplication in $E$. See Theorem 6.1 below, which is a work-out of [9, Th. 3.52] for the spinor Dirac operator. Actually, formula (6.10) is analogous to the formula of Lichnerowicz [52] for the spinor Dirac operator.

**Theorem 6.1** *For the* Spin$^{\mathrm{c}}$*-Dirac operator $D$, acting on sections of $E \otimes L$, we have*

$$D^2 = \Delta + \tfrac{1}{4}\, r^\beta + \tfrac{1}{2}\, \mathrm{c}\left(R^{K^*}\right) + \mathrm{c}\left(R^L\right). \tag{6.10}$$

*In this formula, $r^\beta$ is the scalar curvature of $(M, \beta)$. Furthermore, $\mathrm{c}\left(R^{K^*}\right) \in$ $\mathrm{End}(E) \otimes 1$ and $\mathrm{c}\left(R^L\right)$ are given by*

$$\mathrm{c}(R) = \sum_{j<k} \mathrm{c}\,(\phi_j \cdot \phi_k) \circ R\,(f_j,\, f_k)\,, \tag{6.11}$$

*with $R = R^{K^*}$ and $R = R^L$, respectively. Here $R^{K^*}$, the curvature of the connection in the line bundle $K^*$, is viewed as an $i\,\mathbf{R}$-valued closed two-form in $M$. The curvature operator $R^L$ of $\nabla^L$ is an* $\mathrm{End}(L)$*-valued two-form in $M$. Finally, $f_j$ is any local orthonormal frame in* $\mathrm{T}M$ *and $\phi_j$ is the corresponding dual frame in* $\mathrm{T}^*M$.

**Proof** For any local frame $f_j$ in $\mathrm{T}\,M$ and dual frame $\phi_j$, we get from (5.14) and the fact that the Clifford multiplication is covariantly constant:

$$\begin{aligned} D^2 &= \sum_{j,k} \mathrm{c}\,(\phi_j) \circ \nabla_{f_j} \circ \mathrm{c}\,(\phi_k) \circ \nabla_{f_k} \\ &= \sum_{j,k} \mathrm{c}\,(\phi_j) \circ \mathrm{c}\left(\nabla^{\beta}_{f_j} \phi_k\right) \circ \nabla_{f_k} + \mathrm{c}\,(\phi_j) \circ \mathrm{c}\,(\phi_k) \circ \nabla_{f_j} \circ \nabla_{f_k}. \end{aligned}$$

In this formula, we substitute

$$\nabla^{\beta}_{f_j} \phi_k = \sum_l \langle f_l,\, \nabla^{\beta}_{f_j} \phi_k \rangle\, \phi_l = -\sum_l \langle \nabla^{\beta}_{f_j} f_l,\, \phi_k \rangle\, \phi_l$$

and

$$\sum_k \langle \nabla^{\beta}_{f_j} f_l,\, \phi_k \rangle\, f_k = \nabla^{\beta}_{f_j} f_l.$$

This yields, using also that $\mathrm{c}(\phi_j) \circ \mathrm{c}(\phi_k) = \mathrm{c}(\phi_j \cdot \phi_k)$,

$$D^2 = \sum_{j,k} \mathrm{c}\,(\phi_j \cdot \phi_k) \circ \left(-\nabla_{\nabla^{\beta}_{f_j} f_k} + \nabla_{f_j} \circ \nabla_{f_k}\right).$$

We now add the sum with $j$ and $k$ interchanged, divide by 2, and use

$$\phi_k \cdot \phi_j = -\phi_j \cdot \phi_k - 2\,\beta^{-1}\,(\phi_j,\, \phi_k)\,,$$

cf. (3.5). And we use that the Levi-Civita connection is torsion-free:

$$\nabla^{\beta}_{f_j} f_k - \nabla^{\beta}_{f_k} f_j = [f_j,\, f_k]\,.$$

The result is

$$\begin{aligned} D^2 &= \tfrac{1}{2} \sum_{j,k} \mathrm{c}\,(\phi_j \cdot \phi_k) \circ \left(-\nabla_{[f_j,\, f_k]} + \left[\nabla_{f_j}\, \nabla_{f_k}\right]\right) \\ &\quad - \sum_{j,k} \beta^{-1}\,(\phi_j,\, \phi_k)\, \left(-\nabla_{\nabla^{\beta}_{f_k} f_j} + \nabla_{f_k} \circ \nabla_{f_j}\right) \\ &= \tfrac{1}{2} \sum_{j,k} \mathrm{c}\,(\phi_j \cdot \phi_k) \circ R\,(f_j,\, f_k) + \Delta. \end{aligned} \tag{6.12}$$

Note that, if $\nabla^\beta$ would have had torsion, then the torsion would give rise to additional first order terms.

The curvature operator on $E \otimes L$ splits as follows:

$$R(f_j,\, f_k)\,(\omega \otimes \lambda) = \left(R^E(f_j,\, f_k)\,\omega\right) \otimes \lambda + \omega \otimes \left(R^L(f_j,\, f_k)\,\lambda\right).$$

If we now take the $f_j$ orthonormal, then $\mathrm{c}(\phi_j)$ anticommutes with $\mathrm{c}(\phi_k)$ if $j \neq k$, so we recognize $\mathrm{c}\left(R^L\right)$, given by (6.11) with $R = R^L$, as the sum of the $R^L$-terms.

The definition (3.45) of the covariant differentiation in $E$ implies in view of (6.9) that

$$R^E(f_j,\, f_k) = -\,\mathrm{c}\left(\Omega\left(f_{j,\mathrm{hor}},\, f_{k,\mathrm{hor}}\right)\right). \tag{6.13}$$

The connection in the spin-c frame bundle was defined as the pull-back of the connection in $\Delta^*\,(\mathrm{SOF}\,M \times \mathrm{U}(K^*))$, under the covering map (5.6), defined by the mappings (5.3) and (5.4). Reading backward (6.9), this yields

$$-\,\tau'\,\Omega\left(f_{j,\mathrm{hor}},\, f_{k,\mathrm{hor}}\right) = f^{-1} \circ R^\beta(f_j,\, f_k) \circ f, \tag{6.14}$$

if we identify $\mathbf{R}^{2n}$ with $\left(\mathbf{R}^{2n}\right)^*$. Furthermore, identifying the Lie algebra of $\mathrm{U}(1)$ with $i\,\mathbf{R}$, we get for the scalar part in the Lie algebra

$$-\,\iota'\,\Omega\left(f_{j,\mathrm{hor}},\, f_{k,\mathrm{hor}}\right) = R^{K^*}(f_j,\, f_k). \tag{6.15}$$

We use the isomorphism $f : \mathbf{R}^{2n} \to \mathrm{T}_x\, M$ in order to identify $\mathbf{R}^{2n}$ with $\mathrm{T}_x\, M$. And the identification of $\mathbf{R}^{2n}$ with its dual allows us also to write $f_j = \phi_j$.

Using that $R^\beta_{mljk} = -R^\beta_{lmjk}$, we get

$$\begin{aligned} R^\beta(f_j,\, f_k)\,\zeta &= \sum_{l,\,m} R^\beta_{mljk}\,\zeta_l\, f_m \\ &= \tfrac{1}{2} \sum_{l,\,m} R^\beta_{mljk}\,(\zeta_l\, f_m - \zeta_m\, f_l)\,. \end{aligned}$$

On the other hand, if $j \neq k$, then $f_j \cdot f_k = -f_k \cdot f_j$ in $\mathrm{C}(2n)$, and we get, in view of (4.1),

$$\tau'(f_j \cdot f_k)\,\zeta = 2\,(\zeta_j\, f_k - \zeta_k\, f_j)\,.$$

Comparing the two, we see that

$$R^\beta(f_j, f_k) = \tfrac{1}{4} \sum_{l,m} R^\beta_{mljk}\, \tau'(f_l \cdot f_m).$$

In view of (6.14), we conclude that

$$-\Omega\left(f_{j,\mathrm{hor}}, f_{k,\mathrm{hor}}\right) = \tfrac{1}{4} \sum_{l,m} R^\beta_{mljk}\, f_l \cdot f_m + \lambda,$$

for some $\lambda \in i\,\mathbf{R}$. Since $f_l \cdot f_m \in \mathfrak{spin}(2n)$ if $l \neq m$ (the terms with $l = m$ are equal to zero), we get from (4.11) and (6.15) that $\lambda = \frac{1}{2} R^{K^*}(f_j, f_k)$. In combination with (6.13), we arrive at

$$R^E(f_j, f_k) = \tfrac{1}{4} \sum_{l,m} R^\beta_{mljk}\, \mathrm{c}(f_l \cdot f_m) + \tfrac{1}{2} R^{K^*}(f_j, f_k).$$

This shows that the proof is complete if we have

$$\sum_{jklm} R^\beta_{mljk}\, f_j \cdot f_k \cdot f_l \cdot f_m = 2\, r^\beta \text{ in } \mathrm{C}(2n). \tag{6.16}$$

For the next step, we use that the curvature $R^\beta$ of the Levi-Civita connection satisfies the *Bianchi identity*

$$R^\beta(a, b)\, c + R^\beta(c, a)\, b + R^\beta(b, c)\, a = 0 \tag{6.17}$$

for any vector fields $a$, $b$, $c$. (For a proof, see [9, Proposition 1.26, (3)] or [38, Lemma 12.5].)

If $I = \{j, k, l\}$ is a set of three different indices, then we get, using that $f_p \cdot f_q = -f_q \cdot f_p$ when $p \neq q$,

$$\begin{aligned}
&\sum_{\{j',k',l'\}=I} R^\beta_{ml'j'k'}\, f_{j'} \cdot f_{k'} \cdot f_{l'} \cdot f_m \\
&\sum_{\pi} R^\beta_{m\pi(l)\pi(j)\pi(k)}\, f_{\pi(j)} \cdot f_{\pi(k)} \cdot f_{\pi(l)} \cdot f_m \\
&\sum_{\pi} R^\beta_{m\pi(l)\pi(j)\pi(k)}\, \mathrm{sign}(\pi)\, f_j \cdot f_k \cdot f_l \cdot f_m = 0.
\end{aligned}$$

Here $\pi$ runs over all the permutations of $I$, and we have used that the alternating sum of $R^{\beta}_{mljk}$ over $j,\ k,\ l$ is equal to zero because of (6.17). It follows that (summing only over $j \neq k$ and $m \neq l$):

$$\begin{aligned}
&\sum_{jklm} R^{\beta}_{mljk}\, f_j \cdot f_k \cdot f_l \cdot f_m \\
&= \sum_{jkm} R^{\beta}_{mjjk}\, f_j \cdot f_k \cdot f_j \cdot f_m + R^{\beta}_{mkjk}\, f_j \cdot f_k \cdot f_k \cdot f_m \\
&= -2 \sum_{jkm} R^{\beta}_{mkjk}\, f_j \cdot f_m,
\end{aligned}$$

where in the last identity we used that $f_k^2 = -1$.

Finally, using the Bianchi identity (6.17), one also has

$$\beta\left(R^{\beta}(a,\,b)\,c,\,d\right) = \beta\left(R^{\beta}(c,\,d)\,a,\,b\right), \tag{6.18}$$

for any vector fields $a,\ b,\ c,\ d$. We therefore get that $R^{\beta}_{mkjk} = R^{\beta}_{jkmk}$, so the sum over $j \neq m$ drops out. Using that $f_j^2 = -1$, we arrive at (6.16). □

## 6.2 Dirac Operators on Spinor Bundles

Those readers who are more familiar with Dirac operators $D_{\mathrm{S}}$ on spinor bundles will have remarked that there is a connection between the spin-c Dirac operator $D$ and $D_{\mathrm{S}}$. Let us shortly explain this connection, although we will not use it in the sequel.

A *spin structure* on $M$ is a principal $\mathrm{Spin}(2n)$-bundle $P$ over $M$ such that the associated bundle $P \times_{\mathrm{Spin}(2n)} \mathbf{R}^n$ is isomorphic to $\mathrm{T}^* M$. Here $\mathrm{Spin}(2n)$ acts on $\mathbf{R}^n$ by means of $\tau$. Let $\psi : P \times \mathbf{R}^n \to \mathrm{T}^* M$ denote the mapping which induces the isomorphism. Using the Gram-Schmidt orthogonalization procedure, one can arrange that $\psi$ is an isometry on the fibers. Restricting to the standard basis of $\mathbf{R}^n$, we get a mapping $\psi : P \to \mathrm{SOF}\, M$, which is a double covering, intertwining the action of $\mathrm{Spin}(2n)$ on $P$ with the action of $\mathrm{SO}(2n)$ on $\mathrm{SOF}\, M$ by means of $\tau$. The global existence of such a double covering is equivalent to the vanishing of the second Stiefel-Whitney class of $\mathrm{T}\, M$, which is an element of $\mathrm{H}^2\,(M,\, \mathbf{Z}/2\mathbf{Z})$. (For more details, see [9, Prop. 3.34].) Given a spin structure $P$ on $M$, the *spinor bundle* is defined as the associated bundle

$$S = P \times_{\mathrm{Spin}(2n)} \mathrm{E}(n),$$

where $\mathrm{Spin}(2n)$ acts on $\mathrm{E}(n)$ by means of the Clifford multiplication c. In $P$ we take the connection, which is equal to the pullback of the Levi-Civita connection in $\mathrm{SOF}\, M$, by means of the covering $P \to \mathrm{SOF}\, M$. This leads to a covariant differentiation of sections of $S$, as in (3.45). Then, as in (5.14), one obtains a Dirac operator $D_{\mathrm{S}}$, acting on sections of $S \otimes L$. The one described here is often called *the* Dirac operator on $M$.

In order to compare the bundle $E$ with the bundle $S$, it will be convenient to restrict the structure group to the subgroup $\tilde{\mathrm{U}}(n)$ of $\mathrm{Spin}(2n)$, which is the pre-image of $\mathrm{U}(n)$ in $\mathrm{Spin}(2n)$ under $\tau$ so that $\tau$ induces a double covering from $\tilde{\mathrm{U}}(n)$ onto $\mathrm{U}(n)$. We have

$$\tilde{\mathrm{U}}(n) = \{z\, u \mid z \in \mathrm{U}(1),\ u \in U,\ z^2 \, \mathrm{det}_{\mathbf{C}}\, \tau(u) = 1\}, \tag{6.19}$$

cf. (4.14). Since $\mathrm{c}(U) = 1$ on the zero degree part of $\mathrm{E}(n)$, we have $\mathrm{U}(1) \cap U = \{1\}$, so $m : (z,\, u) \mapsto z\, u$ is an isomorphism from $\mathrm{U}(1) \times U$ onto $\mathrm{U}(1) \cdot U \supset \tilde{\mathrm{U}}(n)$. We write $j : \tilde{\mathrm{U}}(n) \to \mathrm{U}(1)$ for $m^{-1}$, followed by the projection onto the first factor. Note that

$$j(\tilde{u})^2 = 1/\, \mathrm{det}_{\mathbf{C}}\, \tau(\tilde{u}).$$

Let $Q$ denote the pre-image in $P$ of $\mathrm{UF}\, M$ under the covering of $\mathrm{SOF}\, M$ by $P$. Then $Q$ is a double covering of $\mathrm{UF}\, M$, and a principal $\tilde{\mathrm{U}}(n)$-bundle over $M$. With the action $z \mapsto z/j(\tilde{u})$ of $\tilde{u} \in \tilde{\mathrm{U}}(n)$ on $\mathbf{C}$, one gets the complex line bundle

$$K^*_{\frac{1}{2}} = Q \times_{\tilde{\mathrm{U}}(n)} \mathbf{C}$$

over $M$. Projecting $Q$ onto SOF $M$ and using $\tau$ to project $\tilde{\mathrm{U}}(n)$ to $\mathrm{U}(n)$, we see from (6.19) and (5.5) that we would get the bundle $K^*$, if we would have taken $z \mapsto z/\jmath(\tilde{u})^2$ instead of $z \mapsto z/\jmath(\tilde{u})$. That is,

$$K^* = K^*_{\frac{1}{2}} \otimes K^*_{\frac{1}{2}}.$$

This explains the somewhat funny notation.

Now $S \otimes K^*_{\frac{1}{2}}$ is the associated $\tilde{\mathrm{U}}(n)$-bundle, where $\tilde{u} \in \tilde{\mathrm{U}}(n)$ acts on $\mathrm{E}(n)$ by means of $\jmath(\tilde{u})^{-1}\, \mathrm{c}(\tilde{u})$. We have $\tilde{u} = \jmath(\tilde{u})\, u$, where $u \in U$ and $\mathrm{c}(u) = \Lambda\,(\tau(u))$, cf. (4.13). This shows that

$$E = S \otimes K^*_{\frac{1}{2}} \quad \text{and} \quad E \otimes L = S \otimes \left(K^*_{\frac{1}{2}} \otimes L\right). \tag{6.20}$$

Since the curvature of a tensor product bundle is equal to the sum of the curvatures of the factors, the curvature of $K^*_{\frac{1}{2}}$ is equal to $\frac{1}{2}\, R\,(K^*)$. We now can compare the formula in [9, p. 134] for the square of $D_{\mathrm{S}}$ (in which $\mathcal{W}$ is replaced by $L$) with (6.10). The conclusion is that these formulas are equivalent if $M$ has a spin structure. See also [32, pp. 186-190].

## 6.3 The Kähler Case

We now return to the discussion of (6.10). Note that in (6.11), there appears the element

$$\phi_j \cdot \phi_k = \tfrac{1}{2}\,(\phi_j \cdot \phi_k - \phi_k \cdot \phi_j) \in \mathfrak{spin}\,(\mathrm{T}^*_x\, M)\,, \tag{6.21}$$

the Lie algebra of the spin group of $T^*_x M$. It acts on $E_x$ via the Clifford multiplication. Obviously, $\phi_j \cdot \phi_k \in \mathrm{C}^{\leq 2}\,(\mathrm{T}^*_x\, M)$.

If $(M,\, J,\, h)$ is a Kähler manifold, and $L$ is a holomorphic vector bundle over $M$, with a Hermitian structure $h^L$ and a complex linear covariant differentiation $\nabla^L$ which leaves $h^L$ invariant, then $\bar{\partial}^2 = 0$ and $\left(\bar{\partial}^*\right)^2 = 0$, so

$$D^2 = 4\left(\bar{\partial}^*\,\bar{\partial} + \bar{\partial}\bar{\partial}^*\right)$$

leaves the grading of the exterior algebra bundle $E$ of $\mathrm{T}^* M^{(0,1)}$ invariant. We therefore expect a formula for $D^2$ in which all the terms do the same, which does not happen with (6.10) and (6.11).

If $(e_j)_{j=1}^n$ is a local unitary frame, then it is convenient to pass to the complex frame

$$Z_j = e_j - i\, Je_j, \; \bar{Z}_j = e_j + i\, Je_j, \quad 1 \leq j \leq n. \tag{6.22}$$

If the real linear forms $\rho_k$ are determined by the condition that

$$\langle e_j, \rho_k \rangle = \delta_{jk}, \; \langle Je_j, \rho_k \rangle = 0, \tag{6.23}$$

then the complex frame which is dual to (6.22) is given by

$$\zeta_k = \tfrac{1}{2}\left(\rho_k + i\, J\rho_k\right), \quad \bar{\zeta}_k = \tfrac{1}{2}\left(\rho_k - iJ\rho_k\right). \tag{6.24}$$

Note that $\bar{\zeta}_k$ is equal to the previously introduced $\epsilon_k = h\, e_k$. With these notations, we get the following translation of Theorem 6.1.

**Proposition 6.1** *If $(M, J, h)$ is a Kähler manifold and $L$ is a holomorphic vector bundle over $M$, then the spin-c Dirac operator $D$, equal to the Dolbeault-Dirac operator $2\left(\bar{\partial} + \bar{\partial}^*\right)$, satisfies*

$$\begin{aligned} D^2 &= \Delta + \tfrac{1}{2} \sum_{j,k} R^{K^*}\left(Z_j, \bar{Z}_k\right) \mathrm{e}\left(\bar{\zeta}_k\right) \circ \mathrm{i}\left(\bar{Z}_j\right) \\ &\quad - \tfrac{1}{2} \sum_{j} R^L\left(Z_j, \bar{Z}_j\right) + \sum_{j,k} \mathrm{e}\left(\bar{\zeta}_k\right) \circ \mathrm{i}\left(\bar{Z}_j\right) \circ R^L\left(Z_j, \bar{Z}_k\right) \end{aligned} \tag{6.25}$$

*With the notation*

$$\Delta^{(0,\bullet)} := -\sum_j \left(\nabla_{Z_j} \circ \nabla_{\bar{Z}_j} - \nabla_{\nabla^\beta_{Z_j} \bar{Z}_j}\right) = \Delta - \tfrac{1}{2} \sum_j R^{E\otimes L}\left(Z_j, \bar{Z}_j\right), \tag{6.26}$$

*this can be rewritten as the "Bochner-Kodaira formula"*

$$D^2 = \Delta^{(0,\bullet)} + \sum_{j,k} \mathrm{e}\left(\bar{\zeta}_k\right) \circ \mathrm{i}\left(\bar{Z}_j\right) \circ R^{K^*\otimes L}\left(Z_j, \bar{Z}_k\right). \tag{6.27}$$

*Here $\mathrm{e}(\zeta)$ and $\mathrm{i}(Z)$ denote the exterior product from the left with the one-form $\zeta$ and the interior product with the vector field $Z$, respectively.*

**Proof** Since $\bar{\partial}^2 = 0$ and $\left(\bar{\partial}^*\right)^2 = 0$, the curvature operators, apart from being complex linear, now also satisfy

$$R(Z_j, Z_k) = 0, \; R\left(\bar{Z}_j, \bar{Z}_k\right) = 0. \tag{6.28}$$

So the first step in the proof of Theorem 6.1 now yields

$$D^2 - \Delta = \tfrac{1}{2} \sum_{j,k} \left[\mathrm{c}(\zeta_j) \circ \mathrm{c}\left(\bar{\zeta}_k\right) - \mathrm{c}\left(\bar{\zeta}_k\right) \circ \mathrm{c}(\zeta_j)\right] \circ R\left(Z_j, \bar{Z}_k\right).$$

But ${\zeta_j}^{(0,1)} = 0$, $\bar{\zeta}_k^{(0,1)} = \bar{\zeta}_k$, whereas $\beta^{-1}\zeta_j = \tfrac{1}{2}\bar{Z}_j$, $\beta^{-1}\bar{\zeta}_k = \tfrac{1}{2} Z_k$. Here we used the complex bilinear extension of the inner product $\beta$. We have $\mathrm{i}(Z_k) = 0$ on complex antilinear forms, so

$$\mathrm{c}(\zeta_j) = -\tfrac{1}{2}\,\mathrm{i}\left(\bar{Z}_j\right), \; \mathrm{c}\left(\bar{\zeta}_k\right) = 2\,\mathrm{e}\left(\bar{\zeta}_k\right).$$

Using finally that

$$\mathrm{i}\left(\bar{Z}_j\right) \circ \mathrm{e}\left(\bar{\zeta}_k\right) = \delta_{jk} - \mathrm{e}\left(\bar{\zeta}_k\right) \circ \mathrm{i}\left(\bar{Z}_j\right),$$

we arrive at

$$D^2 - \Delta = -\tfrac{1}{2} \sum_j R\left(Z_j, \bar{Z}_j\right) + \sum_{j,k} \mathrm{e}\left(\bar{\zeta}_k\right) \circ \mathrm{i}\left(\bar{Z}_j\right) \circ R\left(Z_j, \bar{Z}_k\right).$$

All terms here leave the grading of the exterior algebra bundle $E$ of $\mathrm{T}^* M^{(0,1)}$ invariant.

In order to determine the $E$-part of these curvature contributions, we observe that $R^E\left(Z_j, \bar{Z}_k\right)$ is a derivation, which on $E^1$ acts as minus the transposed operator of $R^\beta\left(Z_j, \bar{Z}_k\right)$. So

$$R^E\left(Z_j, \bar{Z}_k\right) = -\sum_l \mathrm{e}\left(\bar{\zeta}_l\right) \circ \mathrm{i}\left(R^\beta\left(Z_j, \bar{Z}_k\right) \bar{Z}_l\right).$$

The fact that the Levi-Civita connection is torsion-free implies that

$$R^\beta\left(Z_j, \bar{Z}_k\right) \bar{Z}_l + R^\beta\left(\bar{Z}_l, Z_j\right) \bar{Z}_k + R^\beta\left(\bar{Z}_k, \bar{Z}_l\right) Z_j = 0,$$

cf. (6.17). Using that $R^\beta\left(\bar{Z}_k, \bar{Z}_l\right) = 0$, we get that $R^\beta\left(Z_j, \bar{Z}_k\right)\bar{Z}_l$ *is symmetric under the interchange of* $j$ *and* $k$.

As a consequence,

$$\begin{aligned}
&\sum_{j,k} \mathrm{e}\left(\bar{\zeta}_k\right)\circ \mathrm{i}\left(\bar{Z}_j\right)\circ R^E\left(Z_j, \bar{Z}_k\right)\\
&\quad= -\sum_{jkl} \mathrm{e}\left(\bar{\zeta}_k\right)\circ \mathrm{i}\left(\bar{Z}_j\right)\circ \mathrm{e}\left(\bar{\zeta}_l\right)\circ \mathrm{i}\left(R^\beta\left(Z_j, \bar{Z}_k\right)\bar{Z}_l\right)\\
&\quad= -\sum_{jk} \mathrm{e}\left(\bar{\zeta}_k\right)\circ \mathrm{i}\left(R^\beta\left(Z_j, \bar{Z}_k\right)\bar{Z}_j\right)\\
&\quad= -\sum_{jk} \mathrm{e}\left(\bar{\zeta}_k\right)\circ \mathrm{i}\left(R^\beta\left(Z_j, \bar{Z}_j\right)\bar{Z}_k\right).
\end{aligned}$$

Here we used in the second equation that the terms starting with the composition of $\mathrm{e}\left(\bar{\zeta}_k\right)$ and $\mathrm{e}\left(\bar{\zeta}_l\right)$ drop out because of the symmetry with respect to $k$ and $l$ in the curvature. This symmetry has been used once more in the third equation.

So, as in

$$\sum_j R^E\left(Z_j, \bar{Z}_k\right) = -\sum_{jk} \mathrm{e}\left(\bar{\zeta}_k\right)\circ \mathrm{i}\left(R^\beta\left(Z_j, \bar{Z}_j\right)\bar{Z}_k\right),$$

we get an expression in terms of the operator

$$\sum_j R^\beta\left(Z_j, \bar{Z}_j\right).$$

This derivation of $E$ can be expressed in terms of the curvature of $K^*$. In order to see this, we write

$$\begin{aligned}
R^{K^*}(u, v) &= \mathrm{trace}_{\mathbf{C}}\, R^\beta(u, v) = \sum_j i\,\beta\left(R^\beta(u, v)\, e_j, Je_j\right)\\
&= \sum_j i\,\beta\left(R^\beta\left(e, Je_j\right) u, v\right) = \tfrac{1}{2}\sum_j \beta\left(R^\beta\left(Z_j, \bar{Z}_j\right) u, v\right),
\end{aligned}$$

where in the third equation we used (6.18). This shows that

$$\sum_j \mathrm{i}\left(R^\beta\left(Z_j, \bar{Z}_j\right)\bar{Z}_k\right) = -\sum_l R^{K^*}\left(Z_l, \bar{Z}_k\right)\mathrm{i}\left(\bar{Z}_l\right)$$

on $E^1$. Collecting all terms, we get (6.25) and (6.27). □

Note that the derivation $\mathrm{e}\left(\bar{\zeta}_k\right) \circ \mathrm{i}\left(\bar{Z}_j\right)$ in the exterior algebra $E_x$ of $\mathrm{T}_x^*\, M$ maps $\bar{\zeta}_l$ to $\delta_{jl}\, \bar{\zeta}_k$. Therefore, the operators

$$i\ \mathrm{e}\left(\bar{\zeta}_j\right) \circ \mathrm{i}\left(\bar{Z}_j\right), \tag{6.29}$$

$$\mathrm{e}\left(\bar{\zeta}_k\right) \circ \mathrm{i}\left(\bar{Z}_j\right) - \mathrm{e}\left(\bar{\zeta}_j\right) \circ \mathrm{i}\left(\bar{Z}_k\right), \tag{6.30}$$

$$i\ \mathrm{e}\left(\bar{\zeta}_k\right) \circ \mathrm{i}\left(\bar{Z}_j\right) + i\ \mathrm{e}\left(\bar{\zeta}_j\right) \circ \mathrm{i}\left(\bar{Z}_k\right) \tag{6.31}$$

form a basis of

$$\Lambda' \left(\mathfrak{u}\left(\mathrm{T}_x^*\, M\right)\right) = \mathrm{c}(\mathfrak{u}), \tag{6.32}$$

where $\mathfrak{u}$ is the Lie algebra of the unitary subgroup of $\mathrm{Spin}^{\mathrm{c}}\left(\mathrm{T}_x^*\, M\right)$, defined as in (4.7). So the operators acting on $E$, which appear in the curvature contributions to (6.25) and (6.27), are equal to linear combinations of the basis elements (6.29) — (6.31) of the Lie algebra (6.32) of the unitary group, acting on $E$ as derivations.

**Remark**

All the terms in the Bochner-Kodaira formula (6.27) preserve the grading, which means that, for each $q$, they map sections of $\mathrm{T}^*\, M^{(0,\, q)} \otimes L$ to sections of $\mathrm{T}^*\, M^{(0,\, q)} \otimes L$. And for $q = 0$ we have $D^2 = \Delta^{(0,\, 0)}$, because $\mathrm{i}\left(\bar{Z}_j\right) = 0$ on forms of degree zero.

With arguments as in the proof of Lemma 5.5 one can show that

$$\left(\Delta^{(0,\, \bullet)} \omega,\ \nu\right) = \int_M \sum_j h\left(\nabla_{\bar{Z}_j} \omega,\ \nabla_{\bar{Z}_j} \nu\right)\, \mathrm{d}x, \tag{6.33}$$

which may be compared with (6.3). This may be used to give an alternative definition of $\Delta^{(0,\, \bullet)}$ and it shows that this is a positive operator in the sense that $\left(\Delta^{(0,\, \bullet)} \omega,\ \omega\right) \geq 0$ for every compactly supported $\omega$.

One says that a holomorphic line bundle $L$ over a complex analytic manifold $M$ is *positive*, if there exists a Hermitian connection in $L$, such that $\sigma = \frac{1}{2\pi i}\, R^L$ satifies the condition that the $\beta$ defined by (2.24) is positive definite, that is, it defines a Riemannian structure in $M$. The two-form $\sigma$ is equal to the first Chern form of a connection form $\theta$ in the $\mathrm{U}(1)$-bundle $\mathrm{UF}(L)$, cf. (13.6) and (6.9); and as such it is closed, that is $\mathrm{d}\sigma = 0$. It follows that $M$, provided with the Hermitian structure $h = \beta + i\,\sigma$, is a Kähler manifold.

The Bochner-Kodaira formula (6.27) now can be used to prove the *vanishing theorem of Kodaira* [49], which says that if $L$ is a holomorphic complex line bundle over a compact complex analytic manifold such that the line bundle $K^* \otimes L$ is positive, then

$$\mathrm{H}^q\,(M,\, \mathcal{O}(L)) = 0 \quad \text{for every } q > 0. \tag{6.34}$$

See also [75, Ch. IV, Thm. 2.4].

The idea is to apply Bochner's technique to conclude from $D\omega = 0$ that, if we write $R$ for the sum over $j$ in the right hand side of (6.27),

$$0 = \left(D^2\omega,\, \omega\right) = \left(\Delta^{(0,\,\bullet)}\omega,\, \omega\right) + (R\omega,\, \omega) \geq (R\omega,\, \omega)\,,$$

which in view of the positivity of $R$ then implies that the parts of degree $q > 0$ in $\omega$ vanish. The conclusion (6.34) then follows from the isomorphism of $\mathrm{H}^q\,(M,\, \mathcal{O}(L))$ with the kernel of $D = 2(\bar{\partial} + \bar{\partial}^*)$ in the space of sections of $\mathrm{T}^*\, M^{(0,\,q)} \otimes L$.

# Chapter 7

# The Heat Kernel Method

## 7.1 Traces

The goal of the heat kernel method is to express (2.40) as an integral over the fixed point set $M^\gamma$ in $M$ of the transformation $\gamma$. Here $M^\gamma = M$ if $\gamma$ is the identity. The method is based on the following observations about arbitrary elliptic differential operators $D$, acting on sections of a smooth vector bundle $F$ over a compact manifold $M$, which admits a a direct sum decomposition $F = F^+ \oplus F^-$. In our case, $F = E \otimes L$, with the splitting $F^\pm = E^\pm \otimes L$, and $D$ is the spin-c Dirac operator. For the required facts about trace class operators, see for instance Hörmander [42, Sec. 19.1], or Duistermaat [19].

For the general vector bundle $F$, we write $\Gamma = \Gamma(M, F)$, and $\Gamma^\pm = \Gamma(M, F^\pm)$ for the space of smooth sections of $F$ and $F^\pm$, respectively. It is assumed that $D^+ : \Gamma^+ \to \Gamma^-$ is an elliptic differential operator, so its adjoint, denoted by $D^-$, maps $\Gamma^-$ to $\Gamma^+$. If $D = D^+ \oplus D^-$, then $D$ is a selfadjoint elliptic operator on $\Gamma$, with $D^\pm$ equal to the restriction of $D$ to $\Gamma^\pm$.

Write

$$N^\pm = \ker D^\pm = \Gamma^\pm \cap \ker D$$

for the kernel (null space) of $D^{\pm}$ in in $\Gamma^{\pm}$, and let $C^{\pm}$ be the orthogonal complement of $N^{\pm}$ in $\Gamma^{\pm}$. Then $C^{+}$ is equal to the range of $D^{-}$ and $D^{-}$ is an isomorphism from $C^{-}$ onto $C^{+}$. Similarly, $C^{-} = D^{+}\left(\Gamma^{+}\right)$ and $D^{+}$ is an isomorphism from $C^{+}$ onto $C^{-}$.

Now let $T$ be a trace class operator in $\Gamma$, which commutes with $D$ and also preserves the spaces $\Gamma^{\pm}$. We write $T^{\pm}$ for the trace class operator in $\Gamma^{\pm}$ which is obtained by restricting $T$ to $\Gamma^{\pm}$. The condition $T\,D = D\,T$ then translates into the two conditions $T^{-}\,D^{+} = D^{+}\,T^{+}$ and $T^{+}\,D^{-} = D^{-}\,T^{-}$. It follows that $T^{\pm}$ leaves $N^{\pm}$ and $C^{\pm}$ invariant, and that

$$\operatorname{trace}_{\mathbf{C}} T^{\pm} = \operatorname{trace}_{\mathbf{C}} T^{\pm}|_{N^{\pm}} + \operatorname{trace}_{\mathbf{C}} T^{\pm}|_{C^{\pm}}.$$

However, if we write $D_C^{+} = D^{+}|_{C^{+}}$, then

$$T^{+}|_{C^{+}} = \left(D_C^{+}\right) \circ T^{-}|_{C^{-}} \circ \left(D_C^{+}\right)^{-1},$$

and this shows that

$$\operatorname{trace}_{\mathbf{C}} T^{+}|_{C^{+}} = \operatorname{trace}_{\mathbf{C}} T^{-}|_{C^{-}}.$$

Hence, in the computation of the difference of the trace of $T^{+}$ and $T^{-}$, the contributions coming from $C^{+}$ and $C^{-}$ drop out, and we get the *Lefschetz principle*

$$\operatorname{trace}_{\mathbf{C}} T^{+}|_{N^{+}} - \operatorname{trace}_{\mathbf{C}} T^{-}|_{N^{-}} = \operatorname{trace}_{\mathbf{C}} T^{+} - \operatorname{trace}_{\mathbf{C}} T^{-}. \tag{7.1}$$

Now we assume that $T^{\pm}$ are integral operators with smooth integral kernels $K^{\pm}(x,\, y)$:

$$\left(T^{\pm}u\right)(x) = \int_M K^{\pm}(x,\, y)\, u(y)\, \mathrm{d}y.$$

Here, for each $x \in M$ and $y \in M$, $K^{\pm}(x,\, y)$ is a linear mapping from the fiber $F_y^{\pm}$ over $y$ to the fiber $F_x^{\pm}$ over $x$. And $\mathrm{d}x$ denotes a smooth density in $M$.

The point is, that on a compact manifold an integral operator $K_{\text{op}}$ with smooth kernel $K(x, y)$, acting on sections of a vector bundle $F$, is of trace class, and the trace is equal to the integral

$$\operatorname{trace}_{\mathbf{C}} K_{\text{op}} = \int_M \operatorname{trace}_{\mathbf{C}} K(x, x)\, dx \tag{7.2}$$

of the trace of the integral kernel, over the diagonal in $M \times M$. Note that $K(x, x)$ is a linear mapping from $F_x$ to itself, so it makes sense to take its trace.

In order to prove (7.2), one takes an $L^2$-orthonormal basis $e_k(x)$ of sections of $F$, and writes

$$K(x, y) = \sum_{k,l} (K_{\text{op}}\, e_k,\, e_l) \cdot e_l(x) \otimes e_k(y)^*. \tag{7.3}$$

Here $e_k(y)^*$ denotes the linear form $f \mapsto (f, e_k(y))$ on $F_y^{\pm}$. One obtains (7.3) by checking that the integral operator, with kernel equal to the right hand side, gives the same result as $K_{\text{op}}$ when applied to a smooth section $u$ of $F$. If we then let $u$ approach the Dirac-delta function at $y$, we get (7.3). Now putting $x = y$ in (7.3), taking the trace of the resulting endomorphism of $F_x$, and integrating over $x$, one obtains

$$\int_M \operatorname{trace}_{\mathbf{C}} K(x, x)\, dx = \sum_k (K_{\text{op}}\, e_k,\, e_k) = \operatorname{trace}_{\mathbf{C}} K_{\text{op}}.$$

Inserting (7.2) in (7.1), we get

$$\operatorname{trace}_{\mathbf{C}} T^+|_{N^+} - \operatorname{trace}_{\mathbf{C}} T^-|_{N^-} = \int_M \operatorname{str}_{\mathbf{C}} K(x, x)\, dx, \tag{7.4}$$

where the *supertrace* $\operatorname{str}_{\mathbf{C}} K(x, x)$ of the endomorphism $K(x, x)$ of $F_x = F_x^+ \oplus F_x^-$ is defined by

$$\operatorname{str}_{\mathbf{C}} K(x, x) := \operatorname{trace}_{\mathbf{C}} K^+(x, x) - \operatorname{trace}_{\mathbf{C}} K^-(x, x). \tag{7.5}$$

The operator $\gamma$ in (2.40), which commutes with $D$ and leaves $\Gamma^{\pm}$ invariant, is not an integral operator with smooth kernel. Rather, its integral kernel is a Dirac-delta type of distribution along the graph $x = \gamma(y)$ of $\gamma$. However, if we have an integral operator $T$ with smooth kernel as above, such that $\gamma$ commutes with $T$, then $\gamma \circ T = T \circ \gamma$ is an integral operator with smooth kernel

$$(\gamma \circ T\, u)\,(x) = \int_M \gamma|_{F_{\gamma^{-1}x}}\, K\left(\gamma^{-1}x,\, y\right)\, u(y)\, \mathrm{d}y.$$

If we moreover assume that $T^{\pm}$ *acts as the identity on* $N^{\pm}$, then we get, by replacing $T$ by $\gamma\, T$ in (7.4),

$$\mathrm{trace}_{\mathbf{C}}\, \gamma|_{N^+} - \mathrm{trace}_{\mathbf{C}}\, \gamma|_{N^-} = \int_M \mathrm{str}_{\mathbf{C}} \left(\gamma|_{F_x} \circ K(x,\, \gamma x)\right)\, \mathrm{d}x. \tag{7.6}$$

We have also simplified the notation by writing $y = \gamma x$; it is assumed that $\gamma$ preserves the density in $M$.

## 7.2 The Heat Diffusion Operator

In the heat kernel method, the above program is carried out with

$$T = T(t) = \mathrm{e}^{-tQ}, \quad t > 0, \quad Q = D^2, \tag{7.7}$$

the *heat diffusion operator* generated by the operator $D^2 = D^-\, D^+ \oplus D^+\, D^-$. For each $t > 0$, $T(t)$ is an integral operator with integral kernel $K(t,\, x,\, y)$, which depends smoothly on all the variables $(t,\, x,\, y)$. $T(t)$ can alternatively be defined as the solution of the following ordinary differential equation for $t > 0$, with initial condition for $t \downarrow 0$,

$$\frac{dT(t)}{dt} + Q \circ T(t) = 0, \quad \lim_{t \downarrow 0} T(t) = 1. \tag{7.8}$$

Obviously $\mathrm{e}^{-tQ}$ acts as the identity on the kernel of $Q$, which is equal to the direct sum of the kernel of $D^+$ and the kernel of $D^-$. Also, $\mathrm{e}^{tQ}$ leaves $\Gamma^{\pm}$ invariant. The fact that $\gamma$ commutes with $D$, hence with $Q = D^2$, implies that $\gamma$ commutes with $T(t)$. So all the conditions are satisfied and, with $T = T(t)$, (7.6) becomes

$$\mathrm{trace}_{\mathbf{C}}\, \gamma|_{N^+} - \mathrm{trace}_{\mathbf{C}}\, \gamma|_{N^-} = \int_M \mathrm{str}_{\mathbf{C}} \left(\gamma|_{F_x}\, K(t,\, x,\, \gamma x)\right)\, \mathrm{d}x. \tag{7.9}$$

Note that the left hand side is independent of $t > 0$, so apparently the right hand side is independent of $t$ as well.

The next step is the observation that $K(t, x, y)$ is rapidly decreasing as $t \downarrow 0$ for $x \neq y$, and has an asymptotic expansion for $t \downarrow 0$ in powers of $t$ along the diagonal $x = y$ of the form

$$K(t, x, x) \sim t^{-d/2} \sum_{k=0}^{\infty} t^k K_k(x), \quad t \downarrow 0. \tag{7.10}$$

Here $d$ denotes the dimension of $M$, in our case $d = 2n$. Each of the coefficients $K_k(x)$ is given by a universal rational expression in the derivatives of the coefficients of the operator $P$ at the point $x$, where the order of differentiation and the complexity of the expressions increase with $k$. See Gilkey [32, Sec. 1.7] for a proof in this general context. If we insert this in the right hand side of (7.4), then we get the conclusion that

$$0 = \int_M \mathrm{str}_{\mathbf{C}}\, K_k(x)\, dx$$

if $k \neq d/2$, whereas

$$\mathrm{index}\, D^+ = \dim_{\mathbf{C}} N^+ - \dim_{\mathbf{C}} N^- = \int_M \mathrm{str}_{\mathbf{C}}\, K_{d/2}(x)\, dx. \tag{7.11}$$

In particular, the index of $D^+$ can only be nonzero if the dimension of $M$ is even, say $2n$, and then it is equal to the integral of the supertrace of the $n$-th coefficient in the asymptotic expansion for the heat kernel on the diagonal.

If $\gamma$ is not equal to the identity, then the only contributions to the integral (7.9), which are not rapidly decreasing as $t \downarrow 0$, come from a shrinking region near the set of fixed points of $\gamma$. In the case of nondegenerate isolated fixed points, the asymptotic expansion actually starts with a constant term (as a function of $t$), which can easily be determined, see for instance [32, Lemma 1.8.3]. If the fixed point set $M^\gamma$ of $\gamma$ in $M$ locally is a smooth manifold and the action of $\gamma$ is nondegenerate in the normal bundle, then one gets a conclusion analogous to (7.11), but with the integration over $M$ replaced by an integral over the fixed point set, $d$ replaced by the dimension of $M^\gamma$ and with a suitably modified integrand.

The last step of the heat kernel method is to express the *integrand* $\mathrm{str}_{\mathbf{C}}\, K_{d/2}(x)$ in (7.11) in terms of the geometric data, used in the definition of the operator $D$. For higher dimensions $d = 2n$, this seems at first sight to be an intractable quantity because $K_{d/2}(x)$ is a complicated expression involving derivatives of high order of the geometric data at the point $x$.

We will follow here the method of Berline-Vergne, to explain why the supertrace of $K_n(x)$ is given by an explicit polynomial expression in curvatures (and no derivatives of curvatures). It consists of introducing a principal $G$-bundle $P$ over $M$, a heat kernel $K^P(t,\, p,\, q)$ on $\mathbf{R}_{>0} \times P \times P$, and then writing the heat kernel $K(t,\, x,\, y)$ as an integral over $g \in G$ of $K^P(t,\, p,\, g\,q)\,\rho(g)$, where $p$ and $q$ are in the fiber in $P$ over $x$ and $y$, respectively. See (9.24). For (7.11), we have $x = y$ and we may take $p = q$, but then still $q \neq g\,q$ if $g \neq 1$. This makes that even for (7.11), we will need information about the leading term of the heat kernel expansion away from the diagonal. The required information will be provided by Theorem 8.1, the proof of which is the main goal of Chapter 8.

The factor $g \mapsto \rho(g)$ is a homomorphism from $G$ to the ring of endomorphisms of $E$, in which one has the filtration coming from the Clifford algebra $\mathrm{C}(E)$ via the isomorphism of Lemma 3.1. The supertrace depends only on the highest order part of the endomorphism with respect to the filtration. In Chapter 11 we will see how the factor $\rho$ leads to the desired formula for the supertrace in terms of the curvatures.

If $\gamma \neq 1$, then one gets a local formula by multiplying the supertrace integrand with a test function $\psi$ with support in a small neighborhood of a point of the fixed point set $M^\gamma$, where $M^\gamma$ is a manifold. Integrating, in (7.6), over the fiber $N_x$ through $x \in M^\gamma$ of a $\gamma$-invariant normal fibration $N$ to $M^\gamma$, one is left with an integral over $M^\gamma$. The point is that in the latter integral the integrand converges for $t \downarrow 0$ to $\psi$ times an expression in terms of curvatures and the tangent action of $\gamma$ on $\mathrm{T}_x(N_x)$. So in this case the supertrace integrand converges as $t \downarrow 0$ in distribution sense, with limit equal to a density in the fixed point manifold $M^\gamma$. Furthermore, this density has a natural geometric interpretation.

For these local results, one does not need the manifold $M$ to be compact, or the operator $\mathrm{e}^{-tQ}$ to exist, the only ingredient is the heat kernel expansion, defined as a formal power series. In this sense the local formula is a stronger statement than the integral formula for the virtual character. Another remarkable feature is that even for the global integral formula, the proof uses only analysis and differential geometry, and no arguments from algebraic topology, like $K$-theory, have been used. Admittedly, the proof uses quite a lot of differential geometric constructions, and it depends on ones background, whether one would consider it simpler than the proofs which combine differential geometry with algebraic topology. For an exposition of the latter, see for instance Lawson and Michelsohn [51, Ch. 3].

# Chapter 8

# The Heat Kernel Expansion

## 8.1 The Laplace Operator

Our operator $Q = D^2$, the square of the spin-c Dirac operator, has scalar principal symbol. So for the discussion of the asymptotic expansion of its heat kernel, we may restrict ourselves to the case that $Q$ is a second order differential operator, acting on sections of a complex vector bundle $F$ over a $d$-dimensional Riemannian manifold $(M, \beta)$, with principal symbol given by

$$\sigma_Q(\xi) = \beta^{-1}(\xi, \xi) \cdot 1, \quad \xi \in \mathrm{T}^* M. \tag{8.1}$$

The goal of this chapter is to prove the asymptotic expansion in Theorem 8.1, for the integral kernel of the operator $\mathrm{e}^{-tQ}$, as $t \downarrow 0$. In it, the covariant differentiation $\nabla$ of sections of $F$ is used, which is introduced in Lemma 8.1. The appearance of the geodesic distance $d(x, y)$ in the factor $\mathrm{e}^{-d(x,y)^2/4t}$ is motivated by the equation (8.20) and Lemma 8.3. The quantity $j(x, y)$ in (8.48) is the Jacobian at $x$ of the exponential mapping, centered at $y$, cf. (8.44) and (8.46). The factor $j(x, y)^{-1/2}$ in (8.48) is responsible for the appearance of the Todd class in the formula for the Riemann-Roch number and the Lefschetz number, in Proposition 13.1 and Proposition 13.2, respectively.

If $\nabla$ is a covariant differentiation in $F$, with corresponding Laplace operator $\Delta$ defined by (6.1), then (8.1) is equivalent to the condition that $Q$ and $\Delta$ have the same principal symbol, or that $R = Q - \Delta$ is of order at most one. In general $R$ will actually be of order one. However, we have:

**Lemma 8.1** *There is a unique covariant differentiation $\nabla$ in $F$, such that if $\Delta$ denotes the corresponding Laplace operator, the operator $R = Q - \Delta$ is of order zero.*

**Proof** For convenience of notation, we take for $f_j$ a local orthonormal frame. Then we can write

$$Q - \Delta = \sum_j R_j \nabla_{f_j} + \text{zero order}, \tag{8.2}$$

in which the coefficients $R_j = R_{j,\,x}$ are uniquely determined endomorphisms of $F_x$, depending smoothly on $x$.

Another covariant differentiation $\tilde{\nabla}$ in $F$ is obtained by adding to $\nabla$, at each $x \in M$, a linear mapping

$$\Gamma_x : v \mapsto \Gamma_x(v) : \mathrm{T}_x\, M \longrightarrow \mathrm{End}\,(F_x)\,.$$

That is, for each vector field $v$ in $M$, we have $\tilde{\nabla}_v = \nabla_v + \Gamma(v)$. If $\tilde{\Delta}$ is the Laplace operator defined by $\tilde{\nabla}$, then

$$\tilde{\Delta} = \Delta + 2 \sum_j \Gamma(f_j)\, \nabla_{f_j} + \text{zero order.}$$

So, modulo zero order operators, $Q - \tilde{\Delta}$ is the first order operator with coefficients equal to $R_j - 2\,\Gamma(f_j)$. We get that these vanish, if and only if $\Gamma$ is given by

$$\Gamma(v) = \tfrac{1}{2} \sum_j \beta\,(v,\, f_j)\; R_j \tag{8.3}$$

for each $v \in \mathrm{T}\,M$. □

In classical quantum mechanics, the zero order part $R$ of a Schrödinger operator $Q = \Delta + R$ is called the *potential* if the Laplace operator $\Delta$ is accepted as the kinetic energy Hamiltonian. Since in our application we are quite far from this context, we will not use this terminology here.

If $Q$ is equal to the square of the spin-c Dirac operator $D$, then (6.10) shows that the connection is equal to the one in $E \otimes L$, which was used in the definition (5.14) of $D$. And $R$ is equal to a sum of terms in which curvature operators are combined with Clifford multiplications.

## 8.2 Construction of the Heat Kernel

One says that the section $u(t, x) = u_t(x)$ of the vector bundle $F$ over $\mathbf{R}_{>0} \times M$ satisfies the *heat equation* for the operator $Q$ if

$$\tfrac{\partial u}{\partial t} + Q\, u = 0, \tag{8.4}$$

and one says that $u$ has *initial value* $u_0$ if

$$\lim_{t \downarrow 0} u_t = u_0. \tag{8.5}$$

One can make various choices about the topologies in the space of sections with respect to which one requires (8.5) to hold. In our setting of smooth objects, it is natural to assume convergence with respect to the $\mathrm{C}^\infty$-topology. Then (8.4) implies that also all derivatives of $u_t$ with respect to $t$ converge in the $\mathrm{C}^\infty$-topology as $t \downarrow 0$:

$$\lim_{t \downarrow 0} \tfrac{\partial^k u_t}{\partial t^k} = (-Q)^k\, u_0.$$

If $M$ is compact, then a solution $u = u_t$ of (8.4) and (8.5) exists and is unique. It is given by $u_t = T(t)\, u_0$, with $T(t)$ as in (7.7). In terms of the heat kernel $K(t, x, y)$, this means that

$$u(t, x) = \int_M K(t, x, y)\, u_0(y)\, \mathrm{d}y. \tag{8.6}$$

We will choose the density $\mathrm{d}y$ to be the one for which the orthonormal frames in the tangent spaces with respect to $\beta$ have unit density. Using the existence of a heat kernel for the adjoint of $P$ with respect to some Hermitian structure in $F$, one also has a strong uniqueness result, saying that if $\epsilon > 0$, and $u_t$ and $\tilde{u}_t$ are distributional solutions of (8.4) on $]0,\, \epsilon\,[$, such that $u_t - \tilde{u}_t$ converges to 0 in distribution sense as $t \downarrow 0$, then $u_t = \tilde{u}_t$ for all $0 < t < \epsilon$.

There are various ways of proving the existence of a heat operator $T(t)$. The approach of Gilkey [32, Ch. 1] is to prove first that the spectrum of $Q$ lies in a wedge

$$W = \{\lambda \in \mathbf{C} \mid -\epsilon < \arg(\lambda - c) < \epsilon\}$$

where $\epsilon$ is a small positive constant and $c$ is another constant. Inspired by Cauchy's integral formula, one defines

$$T(t) = \tfrac{1}{2\pi i} \int_{\partial W} \mathrm{e}^{-t\lambda}\, (\lambda\, 1 - Q)^{-1}\, \mathrm{d}\lambda.$$

The proof is completed using the theory of $\lambda$-dependent pseudo-differential operators in order to prove the properties of the resolvent $(\lambda\, 1 - Q)^{-1}$, which give that $T(t)$ is a heat operator for $Q$. These properties of the resolvent then also lead to the asymptotic expansion of the heat kernel $K(t,\, x,\, y)$ for $t \downarrow 0$, which we will aim at next.

If the asymptotic expansion is the main goal, then it is actually more efficient to prove the existence of the heat kernel by first constructing the terms of an expansion, and then show by a perturbation argument that it is asymptotic to an exact heat kernel, as $t \downarrow 0$. This is the approach used by Patodi [63, Sec. 4] and Berline, Getzler and Vergne [9, Secs. 2.3-5]. (For the scalar case, the construction of the expansion goes back at least to Minakshisundaram and Pleijel [60, Sec. 1].) This proof has the advantage that the expansion is a matter of formal power series at points $(t,\, x,\, y)$ such that $t = 0$ and $x = y$, and does not need the compactness of $M$ to be uniquely defined. For the existence of the exact heat kernel however, one needs either compactness of $M$ or growth conditions at infinity.

**Remark** A quite different proof can be given via the solution $C(t)$ of the following wave equation with initial conditions:

$$\frac{d^2C(t)}{dt^2} + Q \circ C(t) = 0, \quad C(0) = I, \; C'(0) = 0. \tag{8.7}$$

The advantage here is that, as a consequence of the finite speed of wave propagation, the integral kernel of $C(t)$ has its support contained in the set of $(x, y) \in M \times M$ such that $d(x, y) \leq |t|$. Therefore its construction for small $|t|$ is a purely local matter, for which we do not need the compactness of $M$. The integral kernel of $C(t)$ is a distribution in $M \times M$, with singularities at the front $d(x, y) = |t|$. More precisely, if we include $t$ in the variables, then the integral kernel is a Fourier integral distribution, see for instance Duistermaat [22, Sec. 5.1].

Now the heat diffusion operator can be expressed in terms of the wave propagator $C(t)$ by means of the formula

$$\mathrm{e}^{-tQ} = \frac{1}{\sqrt{4\pi t}} \int_{-\infty}^{\infty} \mathrm{e}^{-\tau^2/4t}\, C(\tau)\, \mathrm{d}\tau. \tag{8.8}$$

The desired asymptotic expansion of the heat kernel can be proved by substituting the Fourier integral representation of $C(\tau)$ in (8.8) and applying the method of steepest descent to the resulting multiple integral.

It might also be noted that if $Q = D^2$, then

$$C(t) = \cos(t\,D) = \tfrac{1}{2}\left(\mathrm{e}^{it\,D} + \mathrm{e}^{-it\,D}\right).$$

We may also replace $C(tau)$ by $\mathrm{e}^{i\tau\,D}$ in (8.8), because $\mathrm{e}^{-\tau^2/4t}$ is an even function of $\tau$.

## 8.3 The Square of the Geodesic Distance

Since the asymptotic expansion is local near $x = y$, we may work in a local trivialization of the vector bundle and in local coordinates for $x$ in a neighborhood of $y$, where we take $y$ as the origin of the coordinate system. It is natural to expect that an asymptotic expansion of the heat kernel will have a leading term which is equal to the heat kernel for the constant coefficient operator, obtained by replacing the coefficients of $Q$ by their values at the origin, and disregarding the terms of order lower than two. In an orthonormal basis this then is the standard Laplace operator

$$-\sum_{j=1}^{d} \partial^2/\partial {x_j}^2,$$

for which the heat kernel is equal to

$$q(t,\,x,\,y) := (4\pi t)^{-d/2}\,\mathrm{e}^{-\phi(x,\,y)/2t}, \tag{8.9}$$

with

$$\phi(x,\,y) = \tfrac{1}{2}\,\|x-y\|^2 = \tfrac{1}{2}\sum_{j=1}^{d}(x_j-y_j)^2\,. \tag{8.10}$$

We therefore expect an expansion for $K(t,\,x,\,y)$ of the form

$$\tilde{K}(t,\,x,\,y) = q(t,\,x,\,y)\sum_{k=0}^{\infty} t^k\,A_k(x,\,y) \quad A_0(x,\,x) = 1. \tag{8.11}$$

Here $q$ is as in (8.9), where now $\phi$ is a smooth function defined in a neighborhood of the diagonal $x = y$ in $M \times M$, and satisfies, for every $x \in M$:

$$\phi(x,\,x) = 0, \tag{8.12}$$

$$\frac{\partial\phi(x,y)}{\partial y}\Big|_{y=x} = 0, \tag{8.13}$$

$$\frac{\partial^2\phi(x,y)}{\partial y^2}\Big|_{y=x} = \beta_x. \tag{8.14}$$

Note that (8.12) implies that

$$\frac{\partial\phi(x,y)}{\partial x} + \frac{\partial\phi(x,y)}{\partial y} = 0 \quad \text{when} \quad x = y,$$

so (8.13) now implies also that, for every $y \in M$,

$$\frac{\partial\phi(x,y)}{\partial x}\Big|_{x=y} = 0. \tag{8.15}$$

That is, the total derivative $\mathrm{d}\phi$ of $\phi$ vanishes along the diagonal in $M \times M$. The condition (8.13) also gives that the Hessian of $y \mapsto \phi(x, y)$ at $x$ is a well-defined symmetric bilinear form on $\mathrm{T}_x M$. Then (8.14) says that we require that it is equal to the inner product $\beta_x$. From (8.13) and (8.15), we get, *when* $x = y$,

$$\frac{\partial^2 \phi(x,y)}{\partial x\, \partial y} + \frac{\partial^2 \phi(x,y)}{\partial y^2} = 0,$$
$$\frac{\partial^2 \phi(x,y)}{\partial x^2} + \frac{\partial^2 \phi(x,y)}{\partial y\, \partial x} = 0.$$

So, because $\partial/\partial x$ and $\partial/\partial y$ commute, we also have

$$\frac{\partial^2 \phi(x,y)}{\partial x^2}|_{x=y} = \beta_y. \tag{8.16}$$

In (8.11), the right hand side is a formal power series in $t$. For each $j$, the amplitude coefficient $x \mapsto A_j(x, y)$ is a jet (the coordinate-invariant object which in local coordinates corresponds to a formal power series) at $x = y$, with values in the space of linear mappings from $E_y$ to $E_x$. In a local trivialization of the vector bundle $E$, and in local coordinates in the manifold $M$, these are just square matrices, each of the coefficients of which are formal power series in $x - y$ of the form

$$A_k(x, y) \sim \sum_{\alpha} (x - y)^{\alpha}\, A_{k,\alpha}(y). \tag{8.17}$$

Here we use the multi-index notation that $\alpha = (\alpha_j)_{j=1}^{n}$, $\alpha_j \in \mathbf{Z}_{\geq 0}$, and

$$z^{\alpha} = \prod_{j=1}^{n} {z_j}^{\alpha_j}. \tag{8.18}$$

It is clear that for (8.11), it is sufficient to have that also $\phi(x, y)$ only exists as a jet along the diagonal $x = y$.

The reason for the choice $A_0(x, x) = 1$ is that $y \mapsto q(t, x, y)$ converges for $t \downarrow 0$ to the Dirac measure at $x$. It follows that, for every $u_0$, the function $u$ defined by (8.6) satisfies the initial condition (8.5), if and only if $A_0(x,\, x) \equiv 1$. That is, if (8.11) is actually a locally uniform asymptotic expansion, for $t \downarrow 0$ and $y \to x$, of a smooth kernel function $K(t,\, x,\, y)$ with continuous dependence of the coefficients $y \mapsto A_{k,\,\alpha}(y)$ in (8.17).

Already under the weak assumptions (8.12) — (8.14) about the function $\phi$, it can be proved that there is a unique formal power series $\tilde{K}(t,\, x,\, y)$ at $t = 0,\, x = y$, as in (8.11), which, as a formal power series, satisfies the differential equation

$$\frac{\partial \tilde{K}(t,x,y)}{\partial t} + Q_x\, \tilde{K}(t,\, x, y) = 0. \tag{8.19}$$

Here the subscript $x$ in $Q_x$ denotes that $Q$ acts on the power series $x \mapsto \tilde{K}(t,\, x,\, y)$. In order to be able to determine the leading term $A_0(x,\, y)$, of which we will later need the expansion at $x = y$ up to an order which increases with the dimension, we will make a special choice for $\phi$.

In order to find the natural condition for $\phi$, we write

$$\begin{aligned}
\nabla_v\,(q\,A) &= q\left(-\tfrac{v\phi}{2t}\,A + \nabla_v A\right),\\
{\nabla_v}^2\,(q\,A) &= q\left(\left(\tfrac{v\phi}{2t}\right)^2 A - \tfrac{v\phi}{t}\,\nabla_v A - \tfrac{v^2\phi}{2t}\,A + {\nabla_v}^2 A\right),\\
\tfrac{\partial}{\partial t}\,(q\,A) &= q\left(\left(-\tfrac{d}{2t} + \tfrac{\phi}{2t^2}\right) A + \tfrac{\partial A}{\partial t}\right),
\end{aligned}$$

in which all differential operators act with respect to $t$ and $x$.

So, if $f_j$ is an orthonormal frame, we get, after collecting all terms,

$$\left(\tfrac{\partial}{\partial t} + Q_x\right)\,(q\,A) = q\,SA,$$

in which

$$\begin{aligned}
SA &= \tfrac{1}{2t^2}\left(\phi - \tfrac{1}{2}\sum_j (f_j\phi)^2\right)\,A\\
&\quad + \tfrac{1}{2t}\left((-d - \Delta\phi)\,A + 2\sum_j (f_j\phi)\,\nabla_{f_j} A\right)\\
&\quad + \tfrac{\partial A}{\partial t} + Q_x\,A.
\end{aligned} \tag{8.20}$$

In the $t^{-2}$-term in (8.20), there appears the quantity

$$\sum_j (f_j\phi)^2 = \beta^{-1}(\mathrm{d}\phi,\, \mathrm{d}\phi)\,, \tag{8.21}$$

the square of the length of $\mathrm{d}\phi$ with respect to the inner product $\beta^{-1}$ in $\mathrm{T}^* M$. Here the derivative is taken with respect to the variable $x$. It follows from the weak assumptions (8.12), (8.13), (8.14) about $\phi$ that the coefficient

$$\phi - \tfrac{1}{2}\,\beta^{-1}(\mathrm{d}\phi,\, \mathrm{d}\phi) \tag{8.22}$$

of the $t^{-2}$-term in (8.20) vanishes on the diagonal $x = y$. However, one can actually choose the function $\phi$, such that (8.22) vanishes *identically* in a neighborhood of the diagonal. See Lemma 8.3 below. From the point of view of the general theory of differential equations, this is somewhat surprising, because the first order derivatives in (8.22) enter in a quadratic way. So precisely at $x = y$, which is the place where $\mathrm{d}\,\phi = 0$, this leads to a degenerate equation. For the proof of Lemma 8.3, we need some facts about geodesics, which we now briefly recall.

The *geodesics* in $(M,\, \beta)$ are usually defined as the solution curves of the Euler-Lagrange variational equations

$$\frac{dx}{dt} = v, \quad \frac{d}{dt}\frac{\partial L}{\partial v} - \frac{\partial L}{\partial x} = 0, \tag{8.23}$$

in which

$$L(x,\, v) = \tfrac{1}{2}\,\beta_x(v,\, v), \quad x \in M,\, v \in \mathrm{T}_x\, M \tag{8.24}$$

is the kinetic energy, viewed as a real-valued function on the tangent bundle. It is one of the beautiful discoveries of Hamilton, that for quite general Euler-Lagrange equations (8.23), the transformation

$$v \mapsto \xi = \frac{\partial L(x,v)}{\partial v} \tag{8.25}$$

from $\mathrm{T}\, M$ to $\mathrm{T}^*\, M$, mapping velocities to momenta, transforms (8.23) into the Hamitonian system

$$\frac{dx}{dt} = \frac{\partial f(x,\xi)}{\partial \xi}, \quad \frac{d\xi}{dt} = -\frac{\partial f(x,\xi)}{\partial x} \tag{8.26}$$

in $\mathrm{T}^* M$. Here the real-valued function $f$ on $\mathrm{T}^* M$ is defined by

$$f(x,\,\xi) = \langle v,\,\xi\rangle - L(x,\,v), \quad \tfrac{\partial L(x,v)}{\partial v} = \xi, \tag{8.27}$$

the Legendre transform of $L$. In our case,

$$\xi = \beta_x\, v, \quad f(x,\,\xi) = \tfrac{1}{2}\,{\beta_x}^{-1}(\xi,\xi), \tag{8.28}$$

which is the kinetic energy, regarded as a function of the momentum.

The canonical nature of Hamitonian systems (8.26) was explained by Lie, who started by considering the "tautological one-form" $\tau$ on $\mathrm{T}^* M$, defined by

$$\tau_\xi = \xi \circ \mathrm{T}_\xi\,\pi, \quad \xi \in \mathrm{T}^* M. \tag{8.29}$$

Here, as usual $\pi$ denotes the projection from the bundle ($\mathrm{T}^* M$) to its base space ($M$). The canonical two-form $\sigma$ of $\mathrm{T}^* M$ then is defined by

$$\sigma = \mathrm{d}\tau. \tag{8.30}$$

In local coordinates, $\tau$ and $\sigma$ read as

$$\tau = \sum_{j=1}^{d} \xi_j\; \mathrm{d}x_j$$

and

$$\sigma = \sum_{j=1}^{d} \mathrm{d}\xi_j \wedge \mathrm{d}x_j,$$

respectively, from which it is clear that, at each point of $\mathrm{T}^* M$, the two-form $\sigma$ is nondegenerate. In other words, $\sigma$ is a symplectic form in the cotangent bundle $\mathrm{T}^* M$, cf. Section 15.1.

The Hamiltonian vector field $\mathrm{H}_f$ in the right hand sides of (8.26) now is determined by the equation

$$\mathrm{i}\,(\mathrm{H}_f)\,\sigma + \mathrm{d}f = 0. \tag{8.31}$$

Taking the exterior derivative of (8.31), one gets that $\mathcal{L}\,(\mathrm{H}_f)\;\sigma = 0$, which means that $\sigma$ is invariant under the time-$t$-flow $\mathrm{e}^{t\,\mathrm{H}_f}$, generated by (8.26). Actually, the flow of a vector field $v$ leaves $\sigma$ invariant, if and only if the one-form $\mathrm{i}(v)\,\sigma$ is closed, which locally is equivalent to $v = \mathrm{H}_f$ for some function $f$ (and globally if $\mathrm{H}^1(M,\,\mathbf{R}) = 0$.) Taking the inner product in (8.31) with $\mathrm{H}_f$, we also see that $\mathrm{H}_f\;f = 0$, or $f$ is constant along the solution curves of $\mathrm{H}_f$.

Let $G^t$ denote the geodesic flow in $\mathrm{T}\,M$ after time $t$, the flow defined by (8.23). Then, for each $y \in M$ and $v \in \mathrm{T}_y\,M$, we write

$$\exp_y(v) :=\!\!\mapsto x = \pi\left(G^1(v)\right) \tag{8.32}$$

for the position $\gamma(1)$ in $M$ of the geodesic $\gamma(t)$ such that $\gamma(0) = x$ and $\gamma'(0) = v$. The mapping

$$\exp_y : v \mapsto \exp_y(v)$$

from $\mathrm{T}_y\,M$ to $M$ is called the *exponential mapping centered at* $y$. Using the definition (8.32) and the implicit function theorem, one verifies that $\exp_y$ defines a diffeomorphism from an open neighborhood $U$ of $0$ in the vector space $\mathrm{T}_y\,M$ onto an open neighborhood $V$ of $y$ in $M$.

The norm of the vector $\gamma'(t)$ is constant, equal to the norm $\|v\|$ of $v$. Therefore the length of the geodesic $\gamma(t)$, as $t$ runs from $0$ to $1$, is equal to $\|v\|$. For this reason, the *geodesic distance* between $x$ and $y$ in $M$, near to each other, can now be identified with

$$d(x,\,y) = \beta_y(v,\,v)^{1/2}, \tag{8.33}$$

if $x = \exp_y(v)$ and $v \in U \subset \mathrm{T}_y\,M$. It follows in particular that $(x,\,y) \mapsto d(x,\,y)^2$ is a smooth function in an open neighborhood of the diagonal $x = y$ in $M \times M$.

The inverse of $\exp_y$, followed by a linear orthogonal coordinatization of $\mathrm{T}_y\,M$, defines a regular coordinate system for $M$ in $V$, called the *system* *of normal coordinates, centered at $y$.* The property of normal coordinates which we need is that, at a point $v$ in normal coordinates, the standard inner product of a vector in the direction of $v$ with any other vector is equal to their inner product according to the Riemannian structure. More explicitly:

**Lemma 8.2** *Let $y \in M$, $v \in \mathrm{T}_y M$, $x = \exp_y(v)$. Write $\mathrm{E}_v$ for $v$, regarded as a tangent vector to $\mathrm{T}_y M$ at the point $v$. Then we have*

$$\beta_x\left(\mathrm{T}_v\exp_y(\mathrm{E}_v),\, \mathrm{T}_v\exp_y(w)\right) = \beta_y(v,\, w) \tag{8.34}$$

*for every $w \in \mathrm{T}_y M$.*

**Proof** In the cotangent bundle, we have the exponential mapping

$$\epsilon_y := \pi \circ \mathrm{e}^{\mathrm{H}_f}\,|_{\mathrm{T}_y^* M} : \mathrm{T}_y^* M \to M, \tag{8.35}$$

and we have the relation $\exp_y = \epsilon_y \circ \beta_y$ with the exponential mapping in $\mathrm{T}_y M$.

Let E be the vector field in $\mathrm{T}^* M$, defined by

$$\mathrm{i}(\mathrm{E})\,\sigma = \tau. \tag{8.36}$$

This is the radial vector field in the fibers of $\mathrm{T}^* M$, given in local coordinates by

$$\mathrm{E}_\xi = \sum_{j=1}^{d} \xi_j \frac{\partial}{\partial \xi_j}.$$

If $m$ is a number, then Euler's differential equation $\mathrm{E}f = m\, f$ is equivalent to the condition that the function $\xi \mapsto f(x,\, \xi)$ is homogeneous of degree $m$. For this reason, E is called the *Euler vector field* in $\mathrm{T}^* M$. In our case, where $f$ is the kinetic energy function, we have

$$\mathrm{E}\, f = 2\, f. \tag{8.37}$$

Taking the inner product with E in (8.31), we get

$$0 = \mathrm{i}(\mathrm{E})\, \mathrm{i}\,(\mathrm{H}_f)\,\sigma + 2f = -\,\mathrm{i}\,(\mathrm{H}_f)\,\tau + 2f,$$

so

$$\mathcal{L}(\mathrm{H}_f)\,\tau = \mathrm{i}\,(\mathrm{H}_f)\,\sigma + \mathrm{d}(\mathrm{i}\,(\mathrm{H}_f)\,\tau) = -\,\mathrm{d}f + 2\,\mathrm{d}f = \mathrm{d}f,$$

and therefore

$$\left(\mathrm{e}^{t\,\mathrm{H}_f}\right)^*\tau = \tau + t\,\mathrm{d}f.$$

Let us abbreviate the time-one-flow of $\mathrm{H}_f$ by $\Phi$. Using that $\Phi^*\sigma = \sigma$, we get

$$\mathrm{i}\,(\Phi^*\,\mathrm{E})\,\sigma = \Phi^*\,(\mathrm{i}(\mathrm{E})\,\sigma) = \Phi^*\tau = \tau + \mathrm{d}f = \mathrm{i}\,(\mathrm{E} - \mathrm{H}_f)\,\sigma,$$

so the equation $\Phi^*\tau = \tau + \mathrm{d}f$ can equivalently be presented as

$$\Phi^*\,\mathrm{E} = \mathrm{E} - \mathrm{H}_f\,.$$

At $\xi$, the equation $\Phi^*\tau = \tau + \mathrm{d}f$ becomes

$$\Phi(\xi)\circ \mathrm{T}_{\Phi(\xi)}\,\pi \circ \mathrm{T}_\xi\,\Phi = \xi \circ \mathrm{T}_\xi\,\pi + \mathrm{T}_\xi\,f,$$

which, by restriction to $\mathrm{T}_y^*\,M$, yields

$$\Phi(\xi)\circ \mathrm{T}_\xi\,(\epsilon_y) = {\beta_y}^{-1}\xi, \tag{8.38}$$

where the exponential mapping $\epsilon_y$ is the restriction to $\mathrm{T}_y^*\,M$ of $\pi\circ\Phi$. On the other hand, writing $\Phi^*\,\mathrm{E} = \mathrm{E} - \mathrm{H}_f$ at $\Phi(\xi)$, we get

$$\mathrm{T}_\xi\,\Phi\,(\mathrm{E}_\xi) = \mathrm{E}_{\Phi(\xi)} + \mathrm{H}_{f,\,\Phi(\xi)},$$

which, after projection to the base, yields:

$$\mathrm{T}_\xi\,\epsilon_y(\mathrm{E}_\xi) = {\beta_x}^{-1}\,\Phi(\xi). \tag{8.39}$$

We can use (8.39) in order to eliminate the unknown $\Phi(\xi)$ in (8.38) and thereby arrive at

$$\beta_x^{-1}\,(\mathrm{T}_\xi\,\epsilon_y(\mathrm{E}_\xi),\,\mathrm{T}_\xi\,\epsilon_y(\eta)) = {\beta_y}^{-1}(\xi,\,\eta),$$

for $\xi,\,\eta \in \mathrm{T}_y^*\,M$. Returning to the tangent bundle, we get (8.34). □

**Lemma 8.3** *There exists a unique smooth function $\phi(x, y)$ in a neighborhood of the diagonal $x = y$ in $M \times M$, which at the diagonal vanishes of second order, with nondegenerate transversal Hessian, and which satisfies the differential equation*

$$\tfrac{1}{2}\,\beta^{-1}\,(d_x\phi, d_x\phi) = \phi. \tag{8.40}$$

*This function is given by*

$$\phi(x, y) = \tfrac{1}{2}\,d(x, y)^2, \tag{8.41}$$

*in which $d(x, y)$ denotes the geodesic distance between $x$ and $y$ in $M$.*

**Proof** Assuming that $\phi$ is defined by (8.41), we will show that it satisfies the differential equation (8.40). In normal coordinates with the point $y$ as the origin, we have that

$$d(x, y)^2 = \|x\|^2 = \sum_j {x_j}^2,$$

cf. (8.33). If $x \neq 0$, then we take an orthonormal frame $f_j$ at $x$ for $\beta_x$, such that $f_1$ points in the direction of $x$. It follows from (8.34) that $\|f_1\| = 1$, so we have

$$(f_1\phi)\,(x) = \|x\|.$$

On the other hand, (8.34) also yields that if $j > 1$, then $f_j$ is orthogonal to $x$ with respect to the standard inner product in $\mathbf{R}^d$, and hence

$$(f_j\phi)\,(x) = 0 \quad \text{if} \quad j > 1.$$

This proves (8.40).

Now suppose that $\phi$ is as in the first sentence of the lemma. In order to prove its uniqueness, we recall the starting point of the Hamilton-Jacobi theory of a first order (nonlinear) partial differential equation of the form

$$f\,(x, \mathrm{d}\phi(x)) = \phi(x). \tag{8.42}$$

Define the vector field $v = \mathrm{H}_f + \mathrm{E}$ in $\mathrm{T}^* M$ as the sum of the Hamiltonian vector field of $f$ and the Euler vector field. Then the observation is that (8.42) implies that $v$ is tangent to the manifold

$$G = \{(x,\, \mathrm{d}\phi(x)) \mid x \in M\},$$

the graph of $\mathrm{d}\phi$.

This follows by differentiating (8.42). A coordinate-invariant way of expressing the subsequent algebraic manipulations can be obtained by viewing $\omega = \mathrm{d}\phi$ as a mapping from $M$ to $\mathrm{T}^* M$ and writing

$$\omega^* \left(\mathrm{i}(v)\, \sigma\right) = -\omega^* \left(\mathrm{d}f\right) + \omega^* \tau = -\, \mathrm{d}(f \circ \mathrm{d}\phi) + \mathrm{d}\phi = 0.$$

This shows that, at the point $\xi \in G$, $v_\xi$ is in the orthogonal complement $T'$ of $T = \mathrm{T}_\xi\, G$, with respect to the antisymmetric bilinear form $\sigma$. However,

$$\omega^* \sigma = \omega^* \left(\mathrm{d}\tau\right) = \mathrm{d}(\omega^* \tau) = \mathrm{d}(\mathrm{d}\phi) = 0$$

shows that $T \subset T'$, and because

$$\dim T' = \dim T^* M \;-\; \dim T = 2d - d = d = \dim T,$$

we have $T' = T$, so $v_\xi \in T$.

In our case,

$$f\,(x,\, \xi) = \tfrac{1}{2}\, \beta^{-1}\, (\xi,\, \xi)\,,$$

and we get that $v$ is equal to zero at $Z$, the zero section $\xi = 0$ of $\mathrm{T}^* M$. Furthermore, its linearization at $z \in Z$ has $\mathrm{T}_z\, Z$ as its kernel, and is equal to the identity on its range $R_z$, which is $d$-dimensional. As a consequence, each solution curve $\gamma(s)$ of $v$ converges, for $s \to -\infty$, to a point of $Z$. For $z \in Z$, the *unstable manifold* $U_z$ of $z$ is defined as the set of $\gamma(0)$, where $\gamma$ runs over the solution curves such that $\gamma(s) \to z$ as $s \to -\infty$. $U_z$ is a smooth manifold through $z$ with $\mathrm{T}_z\, U_z = R_z$, see for instance Hirsch, Pugh and Shub [43].

In our case, the graph $G$ of $\mathrm{d}\phi$ is a $d$-dimensional manifold, intersecting $Z$ transversally at $z = (0, y)$, and such that $v$ is tangent to $G$. It follows that every backward solution curve of $v$ in $G$ must converge to $z$, so $G \subset U_z$. Since $\dim G = d = \dim U_z$, the conclusion is that $G$ is locally equal to $U_z$. So $\mathrm{d}\phi$ is locally uniquely determined. But then $\phi$ is uniquely determined as well in view of the condition that $\phi(x, y) = 0$ when $x = y$. □

That $\phi(x, y) = \frac{1}{2}\, d(x, y)^2$ satisfies the differential equation (8.40) has been observed by Hadamard [37, (32) on p. 89], with $\beta$ replaced by a pseudo-Riemannian structure, the principal symbol of a hyperbolic partial differential operator $P$. This observation is the basis of Hadamard's construction of a parametrix for $P$. That $-d(x, y)^2/4t$ is the natural exponent in the heat kernel is also indicated by the theorem of Varadhan [72], which says that, for a scalar operator $Q$, the integral kernel $K(t, x, y)$ of $\mathrm{e}^{-tQ}$ satisfies

$$\lim_{t\downarrow 0} t\, \log K(t, x, y) = -d(x, y)^2/4.$$

The remarkable aspect of this theorem is that it holds globally, for all $x, y \in M$.

## 8.4 The Expansion

In the sequel we assume that $\phi$ satisfies (8.40), so that the $t^{-2}$-terms vanish from (8.20). We fix the point $y \in M$ and view all objects in (8.20) as functions of the variable $x \in M$.

**Lemma 8.4** *In normal coordinates, centered at $y$, $\phi$ is given by (8.10). We also have*

$$\mathrm{E} := \sum_{j=1}^{d} (f_j\phi)\, f_j = \sum_{j=1}^{d} v_j\, \frac{\partial}{\partial v_j}, \tag{8.43}$$

*so* E *is equal to the Euler vector field in $\mathbf{R}^d$.*

**Proof** The first statement follows from (8.41) and (8.33).

For the second statement, we argue as in the first paragraph of the proof of Lemma 8.3. Indeed, if at $v$ we take $f_1$ in the direction of $v$, then (8.34) yields that $\|f_1\| = 1$, hence $f_1 = \|v\|^{-1} \cdot v$, so $f_1\phi = \|v\|$ in view of (8.10). On the other hand, it follows from (8.34) also that, for every $j > 1$, $f_j$ is orthogonal to $f_1$ with respect to the standard inner product. Again using (8.10) we get that $f_j\phi = 0$ for $j > 1$. Thus, at $v$, E is equal to $\|v\| \cdot f_1 = v$.
□

For the determination of $\Delta_x\phi$, we introduce the *Jacobian of the exponential mapping*, defined by

$$j_y(v) = \det\left(\mathrm{T}_v(\exp_y)\right). \tag{8.44}$$

Here the determinant is the real determinant of the linear mapping

$$\mathrm{T}_v(\exp_y) : \mathrm{T}_y\, M \to \mathrm{T}_x\, M, \quad x = \exp_y(v).$$

We also have identified the tangent space at $v$ of $\mathrm{T}_y\, M$ with the vector space $\mathrm{T}_y\, M$ itself. The determinant is computed after identifying both $\mathrm{T}_y\, M$ and $\mathrm{T}_x\, M$ with $\mathbf{R}^d$ by means of an orthonormal basis (frame) in them. Since $x$ is close to $y$, we take the frames close to each other, which implies that these have the same orientation. That is, we arrange that $j(v) > 0$.

The definition (8.44) gives that

$$\int_M \psi \, \mathrm{d}x = \int_{\mathrm{T}_y\, M} \left(\exp_y^* \psi\right)\, j_y \, \mathrm{d}v \tag{8.45}$$

for every function $\psi$ with compact support in $V$. The corresponding function $j$ in a neighborhood of the diagonal in $M \times M$ is defined by

$$j(x, y) = j_y(v) \quad \text{if} \quad x = \exp_y(v). \tag{8.46}$$

**Lemma 8.5** *In normal coordinates, centered at $y$, we have*

$$\Delta\phi = -d - \tfrac{\mathrm{E}j}{j}. \tag{8.47}$$

**Proof** Let $\psi$ be a smooth function with support in $V$. Write $\tilde{\psi} = \exp_y^* \psi$. Using (6.3), (8.45), (8.10) and partial integration, we get the following series of identities:

$$\begin{aligned} \int_M (\Delta\phi)\, \psi \, \mathrm{d}x &= \sum_k \int_M (f_k\phi)\, (f_k\psi)\, \mathrm{d}x \\ &= \int_M \mathrm{E}\psi \, \mathrm{dx} = \int_{\mathrm{T}_y M} \sum_j v_j \frac{\partial\tilde{\psi}}{\partial v_j}\, \jmath(v)\, \mathrm{d}v \\ &= -\int_{\mathrm{T}_y M} \tilde{\psi} \left( d\jmath + \sum_k v_k \frac{\partial \jmath}{\partial v_k} \right) \mathrm{d}x = -\int_M \psi \left( d + \frac{\mathrm{E}\jmath}{\jmath} \right) \mathrm{d}x. \end{aligned}$$

Since this holds for arbitrary $\psi$, we get (8.47). □

Collecting all our preparations, we arrive at the following result[1]:

**Theorem 8.1** *Let $\jmath(x,\, y)$ be defined by (8.46) and (8.44). Then, in local coordinates for $x$ near $y$, there is a unique formal power series expansion in $t$ and $x - y$, of the form*

$$\begin{aligned} \tilde{K}(t,\, x,\, y) &= (4\pi t)^{-d/2}\, \mathrm{e}^{-d(x,\, y)^2/4t} \\ &\quad \cdot \jmath(x,\, y)^{-1/2} \sum_{k=0}^{\infty} t^k\, A_k(x,\, y), \end{aligned} \tag{8.48}$$

*with $A_k$ as in (8.17) and $A_0(x,\, x) = 1$, and which satisfies the heat equation $\frac{\partial \tilde{K}}{\partial t} + Q_x \tilde{K} = 0$.*

*Each of the coefficients in the expansion is given by a universal polynomial expression in finitely many derivatives of the coefficients of the Riemannian structure $\beta$ of $M$, the connection $\nabla$ in $F$, and the coefficients of the zero order part of the operator $Q$. It therefore is a smooth function on $M$.*

*The linear mapping $A_0(x,\, y)$ from $F_y$ to $F_x$ is equal to the parallel transport along the geodesic from $y$ to $x$, with $x$ in the infinitesimal neighborhood of $y$.*

---

[1] Cf. [9, Th. 2.30].

*If $M$ is compact, then the heat kernel $K(t, x, y)$ of the operator $Q$ has $\tilde{K}(t, x, y)$ as its asymptotic expansion for $(t, x) \to (0, y)$. The asymptotic expansion can be termwise differentiated, and the remainder estimates, in terms of higher powers of $t$ and $d(x, y)$, are uniform in $y \in M$.*

**Proof** Applying Lemma 8.3, Lemma 8.4, and Lemma 8.5, we see from (8.20), with $A$ replaced by $j^{-1/2} A$ and the whole expression multiplied by $t$, that the equation $\frac{\partial \tilde{K}}{\partial t} + Q_x \tilde{K} = 0$ is equivalent to the condition that $j^{-1/2} A$ is annihilated by the operator

$$t \, \tfrac{\partial}{\partial t} + \tfrac{\mathrm{E} j}{2j} + \nabla_{\mathrm{E}} + t \, Q_x.$$

Since

$$j^{1/2} \, \nabla_{\mathrm{E}} \left( j^{-1/2} \, A \right) = \mathrm{E} A - \tfrac{\mathrm{E} j}{2j} \, A,$$

we see that this amounts to requiring that $A$ is annihilated by the operator

$$t \, \tfrac{\partial}{\partial t} + \nabla_{\mathrm{E}} + t \cdot j^{1/2} \circ Q_x \circ j^{-1/2};$$

or, that the $A_k$ satisfy

$$k \, A_k + \nabla_{\mathrm{E}} \, A_k + j^{1/2} \, Q_x \left( j^{-1/2} \, A_{k-1} \right) = 0. \tag{8.49}$$

Now the form (8.43) of the Euler vector field E implies that, in normal coordinates and in a local trivialization of $F$ which is horizontal at the point $y$, we have

$$\nabla_{\mathrm{E}} = \sum_k v_k \, \tfrac{\partial}{\partial v_k} + \Gamma,$$

where $\Gamma$ is a zero order operator with coefficients which vanish at the origin. It follows that the coefficient of $v^\alpha$ in (8.49) is of the form

$$(k + |\alpha|) \, A_{k, \alpha} + \sum_{|\beta| < |\alpha|} c_{k, \alpha, \beta} \, A_{k, \beta} + \sum_{|\gamma| \leq |\alpha| + 2} d_{k, \alpha, \gamma} \, A_{k-1, \gamma},$$

where each $c_{k,\alpha,\beta}$ and $d_{k,\alpha,\gamma}$ is a universal polynomial expression in the coefficients of $\beta$, $\nabla$ and $P$. Here

$$|\alpha| = \sum_k \alpha_k$$

denotes the total order of the monomial $v \mapsto v^\alpha$.

The condition that the coefficient of $v^\alpha$ in (8.49) is equal to zero determines $A_{k,\alpha}$ in terms of the $A_{k,\beta}$ with $|\beta| < |\alpha|$ and the $A_{k-1,\gamma}$ with $|\gamma| \leq |\alpha| + 2$. Define $(l, \beta) \leq (k, \alpha)$ if $l \leq k$ and $|\beta| \leq |\alpha| + 2(k - l)$. This introduces a partial ordering in the set of indices. The collection of indices smaller that $(k, \alpha)$ is finite, so we have a finite recursion scheme, and $A_{k,\alpha}$ is determined in terms of the $A_{l,\beta}$, for which $(l, \beta)$ is a minimal index with respect to the partial ordering. Since $(0, 0)$ is the only minimal element and $A_{0,0}$ is required to be equal to 1, this leads to the desired determination of the $A_{k,\alpha}$.

The equation (8.49) for $k = 0$ reads that $\nabla_{\mathrm{E}} A_0 = 0$, so $A_0$ is covariantly constant along the solution curves of E. These are the geodesics emanating from $y$. The fact that the parametrization of the geodesics is not by arc length does not change this conclusion.

For the proof of the statements about the heat kernel in the case of a compact manifold $M$, we refer to [9, Sec. 2.4]. □

Given a formal power series expansion $\tilde{K}(t, x, y)$, there exists a smooth function $K(t, x, y)$ on $\mathbf{R}_{>0} \times M \times M$ such that $\tilde{K}(t, x, y)$ is equal to the asymptotic expansion of $K(t, x, y)$, for $t \downarrow 0$ and $d(x, y) \to 0$, and such that $K(t, x, y)$ is rapidly decreasing for $t \downarrow 0$ if $d(x, y)$ remains bounded away from zero. The function $K$ satisfies the heat equation $\partial K/\partial t + Q_x K = 0$ *asymptotically*, that is $\partial K/\partial t + Q_x K$ is flat at $t = 0$, $x = y$, and rapidly decreasing as $t \downarrow 0$ in all of $M \times M$. We will call such a smooth function $K$ an *asymptotic heat kernel*. The uniqueness of the formal power series expansion means that two such kernels $K$ differ by a kernel which is flat at $t = 0$, $x = y$ and is rapidly decreasing as $t \downarrow 0$ in $M \times M$. This leads to an equivalent formulation of Theorem 8.1, with the formal power series replaced by (an equivalence class of) a smooth kernel $K$, which we take equal to the integral kernel of $e^{-tQ}$ if $M$ is compact. This is the form in which we will apply Theorem 8.1 in the next chapters.

# Chapter 9

# The Heat Kernel on a Principal Bundle

## 9.1 Introduction

We now use that the operator $Q = D^2$ acts on sections of the vector bundle $E \otimes L$, where $E$ is the bundle (5.7), associated to the principal $\mathrm{Spin}^{\mathrm{c}}(2n)$-bundle $\mathrm{Spin}^{\mathrm{c}}\,\mathrm{F}\,M$. The covariant differentiation of sections of this bundle is defined by (3.45).

More generally, and also in order to simplify the notation, let $G$ be a compact connected Lie group with Lie algebra $\mathfrak{g}$, $\pi : P \to M$ a principal $G$-bundle over $M$ and $\rho$ a complex representation of $G$ in a finite-dimensional vector space $\mathrm{E}$. We write $E = P \times_G \mathrm{E}$ for the associated vector bundle over $M$. A section $\nu$ of $E \otimes L$ is a mapping, which assigns to each $p \in P$ an element of $\mathrm{E} \otimes L_x$, $x = \pi(p)$, such that

$$\nu(g\,p) = (\rho(g) \otimes 1)\ \omega(p), \quad g \in G. \tag{9.1}$$

In our case $\mathrm{E} = \mathrm{E}(n)$, and $G = \mathrm{Spin}^{\mathrm{c}}(2n)$, acting on $\mathrm{E}(n)$ by means of the Clifford multiplication (4.4).

Let $\tilde{\Gamma} = \Gamma(P, \mathrm{E} \otimes \pi^* L)$ denote the space of *all* mappings $\nu$ from $P$ to $\mathrm{E} \times L$, such that

$$\nu(p) \in \mathrm{E} \otimes L_{\pi(p)}, \quad p \in P.$$

Then $\Gamma = \tilde{\Gamma}^G$ is the subspace of $G$-invariant elements of $\tilde{\Gamma}$. Here $g \in G$ acts on $\tilde{\Gamma}$ by sending $\nu \in \tilde{\Gamma}$ to the mapping $g\,\nu$, defined by

$$(g\,\nu)(p) = \rho(g)\,\nu\left(g^{-1}\,p\right), \quad p \in P. \tag{9.2}$$

In this chapter, we will introduce an operator of the type "Laplace operator plus zero order terms", which acts on $\tilde{\Gamma}$, and is such that its restriction to $\Gamma$ coincides with $Q$, see Lemma 9.3 below. This will lead to an asymptotic formula for the heat kernel of $Q$ along the diagonal in terms of an asymptotic integral over the Lie algebra $\mathfrak{g}$ of $G$. In this formula all the information about the connection in $E$ is translated in terms of the representation $\rho$, see Theorem 9.1. The modifications neeeded in the presence of an automorphism $\gamma$ are treated in Chapter 10; the result in that case is presented in Theorem 10.1.

## 9.2 The Laplace Operator on $P$

We assume, as usual, that $\beta$ is a Riemannian structure in $M$ and that we have a $G$-invariant connection in $P$, defined by a $\mathfrak{g}$-valued connection one-form $\theta$, and *horizontal subspaces* $H_p = \ker \theta_p$. The restriction of $\theta_p$ to the *vertical space* $V_p$, the tangent space $\ker \mathrm{T}_p\,\pi$ of the fiber, is equal to the inverse of the linear isomorphism $X \mapsto X_{P,p}$ from $\mathfrak{g}$ onto $V_p$. Here $X_P$ denotes the vector field in $P$ which yields the infinitesimal action of $X$ in $P$.

This leads to the covariant differentiation of sections of $E$ as in (3.45). We also assume that $L$ is provided with a Hermitian connection; the corresponding covariant differentiation of sections of $L$ is denoted by $\nabla^L$. Put together, the covariant differentiation $\nabla$ of sections of $E \otimes L$ is determined by (5.13). Now $\nabla_v$ can also be viewed as an operator in $\tilde{\Gamma}$, and as such it leaves the subspace $\Gamma$ invariant because of the $G$-invariance of the connection in $P$. The definition (6.1) leads to an operator $\tilde{\Delta}^M$ on $\tilde{\Gamma}$, which on $\Gamma$ acts as the Laplace operator $\Delta^M$ of $M$.

However, as an operator acting on the space $\tilde{\Gamma}$ of all sections in $P$ of $\mathrm{E} \otimes L$, $\tilde{\Delta}^M$ is not equal to the Laplace operator of any Riemannian structure on $P$ (if $\dim G > 0$), since the second order derivatives in the vertical directions, tangent to the $G$-orbits, are missing.

In order to remedy this, we introduce an $\mathrm{Ad}\, G$-invariant inner product $\beta^{\mathfrak{g}}$ in $\mathfrak{g}$. (For each $g \in G$, $\mathrm{Ad}\, g$ denotes the action of conjugation by means of $g$ in the Lie algebra $\mathfrak{g}$.) With it, we define the Riemannian structure $\beta^P$ in $P$ by

$$\beta_p^P(v,\, w) = \beta_{\pi(x)}^M\left(\mathrm{T}_p\, \pi(v),\, \mathrm{T}_p\, \pi(w)\right) + \beta^{\mathfrak{g}}\left(\theta_p v,\, \theta_p w\right), \tag{9.3}$$

for each $p \in P$ and $v,\, w \in \mathrm{T}_p\, P$. This is equivalent to requiring that $H_p$ is orthogonal to $V_p$, that the inner product coincides on $H_p$ with the horizontal lift of $\beta_x^M$, and on $V_p$ with $\beta^{\mathfrak{g}}$, via the infinitesimal action.

In order to determine the Laplace operator on $P$ with respect to $\beta^P$, as in (6.1), we need to know the covariant derivative of vector fields, defined by the Levi-Civita connection of $\beta^P$. In order to simplify the notation, we will from now on denote all inner products in this chapter by round brackets; so $(v,\, w)$ is the inner product of $v$ and $w$, with respect to the inner product which has been defined in the vector space to which $v$ and $w$ belong.

Let $\Omega$ denote the curvature form of the connection in $P$, a horizontal $\mathfrak{g}$-valued two-form in $P$. For each $X \in \mathfrak{g}$ and $p \in P$, we will use the somewhat drastic abbreviation

$$X \cdot \Omega_p : H_p \to H_p$$

for the endomorphism of the horizontal spaces $H_p$, which is defined by

$$((X \cdot \Omega_p)(v),\, w) = (X,\, \Omega_p(v,\, w)), \qquad v,\, w \in H_p. \tag{9.4}$$

The following lemma is the same as [9, Lemma 5.2]; unfortunately I ended up with quite the opposite notation. The different sign in (9.5) is caused by the fact that here the action is defined to be a homomorphism from $G$ to the diffeomorphism group of $P$, in contrast with the usual convention of writing it as a "right action", which is an anti-homomorphism. However, I did not dare to also change the sign in the convential Lie bracket of vector fields, which is minus the one which one would get if the vector fields would be viewed as the Lie algebra of the group of diffeomorphisms. Sometimes, differential geometry is a true mine-field of notation and sign problems.

**Lemma 9.1** *Let $u$ and $v$ be vector fields in $M$ and let $X$, $Y \in \mathfrak{g}$. Then*

$$\nabla_{X_P} Y_P = -\tfrac{1}{2}\, [X,\, Y]_P = \tfrac{1}{2}\, (\mathrm{ad}Y(X))_P\,, \tag{9.5}$$

$$\nabla_{X_P} v_{\mathrm{hor}} = \tfrac{1}{2}\, (X \cdot \Omega)\, v_{\mathrm{hor}} = \nabla_{v_{\mathrm{hor}}} X_P, \tag{9.6}$$

$$\nabla_{u_{\mathrm{hor}}} v_{\mathrm{hor}} = (\nabla_u v)_{\mathrm{hor}} - \tfrac{1}{2}\, \Omega\, (u_{\mathrm{hor}},\, v_{\mathrm{hor}})_P\,. \tag{9.7}$$

**Proof** In any Riemannian manifold, the covariant differentiation $\nabla$ of the Levi-Civita connection is given by

$$\begin{aligned} 2\, (\nabla_u v,\, w) \;&=\; u\, (v,\, w) + v\, (w,\, u) - w\, (u,\, v) \\ &\quad + ([u,\, v],\, w) - ([v,\, w],\, u) + ([w,\, u],\, v)\,, \end{aligned} \tag{9.8}$$

cf. Helgason [38, proof of Th. 9.1]. In (9.8), one recognizes the definition in terms of Christoffel symbols, if, in local coordinates, $u$, $v$ and $w$ are constant vector fields in the direction of the base vectors. One has to insert the Lie brackets in order to get the correct formula for arbitrary vector fields.

Using that $(X_P,\, Y_P) = (X,\, Y)$ is constant, that $[X_P,\, Y_P] = -[X,\, Y]_P$, and that

$$([Z,\, X],\, Y) + (X,\, [Z,\, Y]) = 0$$

since the inner product in $\mathfrak{g}$ is $\mathrm{Ad}\, G$-invariant, we get that

$$2\, (\nabla_{X_P} Y_P,\, Z_P) = -\, ([X,\, Y],\, Z)\,. \tag{9.9}$$

On the other hand, using also that $(X_P, v_{\text{hor}}) = 0$ and that the vector fields $X_P$ and $v_{\text{hor}}$ in $P$ commute, we have

$$2\left(\nabla_{X_P} Y_P,\, v_{\text{hor}}\right) = 0. \tag{9.10}$$

Combined with (9.9), we get (9.5).

With the same arguments as for (9.10), or using the invariance of $\beta^P$ under $\nabla$, we get

$$2\left(\nabla_{X_P} v_{\text{hor}},\, Z_P\right) = 0. \tag{9.11}$$

And, using (6.8):

$$2\left(\nabla_{X_P} v_{\text{hor}},\, w_{\text{hor}}\right) = \left(\Omega\left(v_{\text{hor}},\, w_{\text{hor}}\right),\, X\right). \tag{9.12}$$

Together with (9.11) and the definition (9.4), this yields (9.6). Note that $\nabla_{X_P} v_{\text{hor}} = \nabla_{v_{\text{hor}}} X_P$, because $X_P$ commutes with $v_{\text{hor}}$ and the connection $\nabla = \nabla^P$ is torsion-free.

There are no new arguments involved in checking that

$$2\left(\nabla_{u_{\text{hor}}} v_{\text{hor}},\, Z_P\right) = -\left(\Omega\left(u_{\text{hor}},\, v_{\text{hor}}\right),\, Z\right). \tag{9.13}$$

And using that $[u_{\text{hor}}, v_{\text{hor}}]$ has horizontal part equal to $[u, v]_{\text{hor}}$, we get

$$2\left(\nabla_{u_{\text{hor}}} v_{\text{hor}},\, w_{\text{hor}}\right) = 2\left(\nabla_u v,\, w\right). \tag{9.14}$$

Together with (9.13), this leads to (9.7). □

The *Casimir operator*, defined by the inner product in $\mathfrak{g}$ and the representation $\rho$, is the endomorphism Cas of E, defined by

$$\text{Cas} = \sum_k \rho'(e_k)^2, \tag{9.15}$$

if $e_k = e_k^{\mathfrak{g}}$ is an orthonormal basis in $\mathfrak{g}$. The right hand side in (9.15) does not depend on the choice of the orthonormal basis $e_k$, and the operator Cas commutes with all $\rho(g)$, $g \in G$. The corresponding operator Cas, acting on sections of $\mathrm{E} \otimes \pi^* L$, is determined by

$$\operatorname{Cas}(\omega \otimes \lambda)(p) = \operatorname{Cas}(\omega(p)) \otimes \lambda(p), \tag{9.16}$$

if $\omega : P \to \mathrm{E}$ and $\lambda \in \Gamma(P, \pi^* L)$.

**Lemma 9.2** *The Laplace operator $\Delta^M$ of $M$ is equal to the restriction to $\Gamma = \tilde{\Gamma}^G$, of the operator $\Delta^P + \operatorname{Cas}$ in $\tilde{\Gamma}$. The operators $\Delta^P$ and $\operatorname{Cas}$ in $\tilde{\Gamma}$ commute and each of them leaves $\Gamma$ invariant.*

**Proof** If $f_j$ is a local orthonormal frame in $\mathrm{T}\,M$ and $e_k$ is an orthonormal basis in $\mathfrak{g}$, then the $f_{j,\,\mathrm{hor}}$ and $e_{k,\,P}$ together form a local orthonormal frame in $\mathrm{T}\,P$. Moreover, we see from (9.7) that

$$\nabla_{f_{j,\,\mathrm{hor}}} f_{j,\,\mathrm{hor}} = \left(\nabla_{f_j} f_j\right)_{\mathrm{hor}},$$

and from (9.5) that $\nabla_{e_{k,\,P}} e_{k,\,P} = 0$. This shows that

$$\Delta^P = \tilde{\Delta}^M - \sum_k \left(\nabla_{e_{k,\,P}}\right)^2.$$

On the other hand, locally any $\nu \in \Gamma$ can be written as a finite sum of elements $\omega \otimes \lambda$, where for each $p \in P$, $g \in G$ we have

$$\omega(g\,p) = \rho(g)\,\omega(p)$$

in E and

$$\lambda(g\,p) = \lambda(p)$$

in $L_x$, $x = \pi(p)$.

Differentiating with respect to $g$, at $g = 1$ and in the direction of $X \in \mathfrak{g}$, we get that

$$(X_P \nu)(p) = \rho'(X)\,\nu(p).$$

Again replacing $p$ by $g\,p$ and differentiating with respect to $g$, at $g = 1$ and in the direction of $X$, we get that

$$(X_P)^2\,\nu = \rho'(X)^2\,\nu.$$

The proof is completed by replacing $X$ by $e_k$ and summing over $k$.

Note that in general $\rho'(X)\,\nu$ will not belong to $\Gamma$, but Cas $\nu \in \Gamma$ because Cas commutes with all $\rho(g)$, $g \in G$. □

## 9.3 The Zero Order Term

We now abstract from (6.10) the properties of the zero order term $R = Q - \Delta^M$, which will be used. At each $x \in M$, each term in (6.11) is of the form $\mathrm{c}(a_x) \otimes B_x$, where $B_x \in \mathrm{End}\,(L_x)$ depends smoothly on $x$ and $a_x$ is the element (6.21) in the Lie algebra of the subgroup $\mathrm{Spin}\,(\mathrm{T}_x^*\,M)$ of $\mathrm{Spin}^{\mathrm{c}}\,(\mathrm{T}_x^*\,M)$.

In order to translate this into the framework of the principal bundle $P$, in our case $\mathrm{Spin}^{\mathrm{c}}\,\mathrm{F}\,M$, we view $f_x$ as an isomorphism from $\mathbf{R}^{2n}$ onto $\mathrm{T}_x\,M$, so its transposed $f_x' : \mathrm{T}_x^*\,M \to \left(\mathbf{R}^{2n}\right)^* \simeq \mathbf{R}^{2n}$ induces an isomorphism $f_x'$ from $\mathrm{C}\,(\mathrm{T}_x^*\,M)$ onto $\mathrm{C}(2n)$. If we write, for lack of a better notation at this point, $\psi_j$ for the standard basis in $\mathbf{R}^{2n}$, we get that

$$f_x'(a_x) = \psi_{jk} := \psi_j \cdot \psi_k \in \mathfrak{g}.$$

Now assume that $e$ is a local unitary frame and that $f = f(e)$, as in (3.44). Then, viewing $e_x$ as an isomorphism from $\mathbf{C}^n$ onto $\mathrm{T}_x\,M$, the pullback operator $e_x^*$ defines an isomorphism from $E_x$ onto $\mathrm{E}(n)$. The definition of the Clifford multiplication (4.4) was arranged in such a way that

$$e^*\left(\mathrm{c}(a)\,\omega\right) = \mathrm{c}\left(f'(a)\right)\,e^*(\omega).$$

In (5.7), $E_x$ was described as the space of $\mathrm{E}(n)$-valued functions $\omega$ on $\mathrm{UF}_x\,M \times \mathrm{Spin}^{\mathrm{c}}(2n)$, such that, for each $e \in \mathrm{UF}_x\,M$, $A \in \mathrm{U}(n)$ and $r,\ s \in$

$\mathrm{Spin}^{\mathrm{c}}(2n)$

$$\omega\left(e \circ A',\, \mathrm{c}^{-1}(\Lambda A) \cdot r \cdot s^{-1}\right) = \mathrm{c}(s)\, \omega\,(e,\, r)\,.$$

And $\omega_x \in E_x$ is identified with

$$\omega : (e,\, r) \mapsto \mathrm{c}(r)^{-1}\, e^*(\omega).$$

Putting these things together, we see that, in the framework of $\mathrm{E}(n) \otimes L_x$-valued functions,

$$\begin{aligned} &\mathrm{c}(R)\,(\omega \otimes \lambda_x)\,(e,\, r) \\ &\quad = \sum_{j<k} \mathrm{c}\left(r^{-1} \cdot \psi_{jk} \cdot r\right)\, \omega(e,\, r) \otimes R_x\,(f(e)_j,\, f(e)_k)\, \lambda_x. \end{aligned} \tag{9.17}$$

Moving over the coefficients of

$$\mathrm{Ad}\, r^{-1} : \psi \mapsto r^{-1} \cdot \psi \cdot r : \mathfrak{g} \to \mathfrak{g}$$

to the $\mathrm{End}(L_x)$-factor, we see that, in terms of the principal $G$-bundle $P$, $\mathrm{c}(R)$ can be written as

$$R^1(p) = \sum_k \rho'(e_k^{\mathfrak{g}}) \otimes R_k^1(p), \tag{9.18}$$

in which $e_k^{\mathfrak{g}}$ denotes an orthonormal basis in $\mathfrak{g}$ and $R_k^1(p) \in \mathrm{End}(L_x)$, $x = \pi(p)$. In our case, $\rho$ and $\rho'$ is equal to the restriction of c to $G = \mathrm{Spin}^{\mathrm{c}}(2n)$ and $\mathfrak{g} = \mathfrak{spin}^{\mathrm{c}}(2n)$, respectively. Both $G$ and $\mathfrak{g}$ are subsets of $\mathrm{C}(2n)$ and $\mathrm{c} : \mathrm{C}(2n) \to \mathrm{End}\,(\mathrm{E}(n))$ is a linear map. Moreover, $R^1$ enjoys the invariance property that

$$\sum_k e_k^{\mathfrak{g}} \otimes R_k^1(g\,p) = \sum_l\, (\mathrm{Ad}\, g)\, (e_l^{\mathfrak{g}}) \otimes R_l^1(p), \quad g \in G. \tag{9.19}$$

The condition (9.19) implies that $R^1$ commutes with the $G$-action in $\tilde{\Gamma}$, which in turn implies that $R^1(\Gamma) \subset \Gamma$. The space of the smooth $p \mapsto R^1(p)$ in (9.18) which satisfy (9.19) will be denoted by

$$\mathcal{R}^1 = \left(\mathfrak{g} \otimes \Gamma\,(P,\, \pi^*\, \mathrm{End}(L))\right)^G. \tag{9.20}$$

**Lemma 9.3** *Let $Q = \Delta^M + R^1 + R^0$, where $R^1 \in \mathcal{R}^1$, cf. (9.18) — (9.20), and $R^0 \in \Gamma(M, \mathrm{End}(L))$. Let $\tilde{\nabla}$ be the covariant differentiation of sections of $\mathrm{E} \otimes \pi^* L$ over $P$, defined by*

$$\tilde{\nabla}_v = \nabla_v^L + 1 \otimes \tfrac{1}{2} \sum_k (v,\, e_k^{\mathfrak{g}})\, R_k^1, \tag{9.21}$$

*and let $\tilde{\Delta}^P$ be the corresponding Laplace operator. Then there exists a zero order operator of the form*

$$\tilde{R}^0 \in \Gamma(P,\, \pi^*\mathrm{End}(L)),$$

*such that $Q$ is equal to the restriction to $\Gamma = \tilde{\Gamma}^G$, of the operator $\tilde{Q} + \mathrm{Cas}$, where*

$$\tilde{Q} = \tilde{\Delta}^P + 1 \otimes \tilde{R}^0. \tag{9.22}$$

*Moreover, $\tilde{Q}$ commutes with the action of $G$ and of* Cas *in $\tilde{\Gamma}$.*

**Proof** The operator $R^1$ in (9.18) acts on $\Gamma = \tilde{\Gamma}^G$ as the first order differential operator

$$\tilde{R}^1 := \sum_k e_k^{\mathfrak{g}} \otimes R_k^1.$$

Applying the proof of Lemma 8.1, and comparing (9.21) with (8.3), we see that

$$\Delta^P + \tilde{R}^1 = \tilde{\Delta}^P + S,$$

where $S \in \Gamma(P,\, \pi^*\,\mathrm{End}(L))$. In combination with Lemma 9.2, we get that $Q$ is equal to the restriction to $\Gamma$ of

$$\Delta^P + R^1 + R^0 + \mathrm{Cas} = \tilde{\Delta}^P + S + R^0 + \mathrm{Cas}\,.$$

This proves the first statement, with $\tilde{R}^0 = S + R^0$.

For the last statement, we observe that (9.19) implies that also the newly introduced operator $\tilde{R}^1$ commutes with the action of $G$. And $\tilde{R}^1$ obviously commutes with Cas. □

## 9.4 The Heat Kernel

Now let $K^P(t, p, q)$, which for each $t > 0$ and $p, q \in P$ is a linear map from $\mathrm{E} \otimes L_{\pi(q)}$ to $\mathrm{E} \otimes L_{\pi(p)}$, denote the integral kernel of the heat diffusion operator $\mathrm{e}^{-t\tilde{Q}}$, where $\tilde{Q}$ is as in (9.22). If $M$ is not compact, then we take for $K^P$ an asymptotic heat kernel, as discussed after Theorem 8.1.

In order to be able to compare the integral kernel $K^M(t, x, y)$ of $\mathrm{e}^{-tQ}$ with $K^P$, we begin by observing that a linear mapping $K(x, y)$ from $E_y \otimes L_y$ to $E_x \otimes L_x$ corresponds, if $x = \pi(p)$ and $y = \pi(q)$, to a linear mapping $K(p, q)$ from $\mathrm{E} \otimes L_y$ to $\mathrm{E} \otimes L_x$, such that

$$K(g p, h q) = \rho(g) \circ K(p, q) \circ \rho(h)^{-1}, \quad g, h \in G. \tag{9.23}$$

The $h$-behavior is necessary and sufficient in order to get a uniquely defined element when applied to an element $\nu$ of $\Gamma = \tilde{\Gamma}^G$, and the $g$-behavior in order to get an element of $\Gamma$ when $K_{\mathrm{op}}$ has been applied.

With this identification, we can now formulate the following result.

**Lemma 9.4** *If $K^P(t, p, q)$ denotes the (asymptotic) integral kernel of $\mathrm{e}^{-t\tilde{Q}}$, with $\tilde{Q}$ as in (9.22), then the integral kernel $K^M(t, x, y)$ of $\mathrm{e}^{-tQ}$ is given by*

$$K^M(t, p, q) = \mathrm{e}^{-t\,\mathrm{Cas}} \int_G K^P(t, p, g q) \circ \rho(g)\, \mathrm{d}g. \tag{9.24}$$

**Proof** We apply Lemma 9.3. The fact that Cas commutes with $\tilde{Q}$ and that both operators leave $\Gamma$ invariant now implies that

$$\mathrm{e}^{-tQ} = \mathrm{e}^{-t\,\mathrm{Cas}} \circ \mathrm{e}^{-t\tilde{Q}} \quad \text{on} \quad \Gamma.$$

Writing this in terms of the integral kernels, and using that

$$\int_P \phi(p)\, \mathrm{d}p = \int_M \left( \int_G \phi(g p)\, \mathrm{d}g \right) \mathrm{d}(Gp), \tag{9.25}$$

we get, for any $\nu \in \Gamma, p \in P, x = \pi(p)$,

$$\int_M K^M(t,\,x,\,y)\,\nu(y)\;\mathrm{d}y = \left(\mathrm{e}^{-tQ}\,\nu\right)(x) = \mathrm{e}^{-t\,\mathrm{Cas}}\left(\mathrm{e}^{-t\tilde{Q}}\,\nu\right)(p)$$
$$= \mathrm{e}^{-t\,\mathrm{Cas}}\int_M\left(\int_G K^P(t,\,p,\,g\,q)\,\nu(g\,q)\;\mathrm{d}g\right)\,\mathrm{d}(Gq)$$
$$= \mathrm{e}^{-t\,\mathrm{Cas}}\int_M\left(\int_G K^P(t,\,p,\,g\,q)\circ\rho(g)\,\nu(q)\;\mathrm{d}g\right)\,\mathrm{d}(Gq).$$

Comparing the first and the last term, we get (9.24). That is, after we have verified that the right hand side in (9.24) has the required property (9.23) of being an integral kernel for an operator on $M$.

In order to see this, we use that the $G$-action in $\tilde{\Gamma}$ commutes with $\tilde{Q}$. This makes that the $G$-action commutes with $\mathrm{e}^{-t\tilde{Q}}$, which in turn means for its integral kernel that

$$K^P(t,\,g\,p,\,g\,q) = \rho(g)\circ K^P(t,\,p,\,q)\circ\rho(g)^{-1},\quad g\in G.$$

Indeed, using (9.2) and the $G$-invariance of the density in $P$, we get for an integral operator $A$ with integral kernel $K$ that

$$(g\,A\nu)(p) = \rho(g)\,(A\nu)(g^{-1}\,p) = \rho(g)\int_P K(g^{-1}\,p,\,q)\,\nu(q)\;\mathrm{d}q,$$

whereas on the other hand

$$(A(g\,\nu))\,(p) = \int_P K(p,\,q)\circ\rho(g)\,\nu(g^{-1}\,q)\;\mathrm{d}q = \int_P K(p,\,g\,p)\circ\rho(g)\,\nu(q)\;\mathrm{d}q.$$

We now replace $p$ and $q$ in (9.24) by $g_1\,p$ and $g_2\,q$, respectively. Then we make the substitution of variables

$$g = g_1\,\tilde{g}\,{g_2}^{-1}.$$

Using the invariance of the density in $G$ under left- and right-multiplications, and the fact that Cas commutes with $\rho(G)$, we get that

$$K^M(t,\,g_1\,p,\,g_2\,q) = \rho(g_1)\circ K^M(t,\,p,\,q)\circ\rho(g_1)^{-1}\circ\rho(g_1)\circ\rho\left({g_2}^{-1}\right).$$

This implies the property (9.23) for the right hand side in (9.24). □

## 9.5 The Expansion

The next step is to substitute the expansion (8.48), with $Q$ replaced by the operator $\tilde{Q}$ of (9.22), in the right hand side of (9.24). If we are aiming at (7.4), then we may put $q = p$, and we have to determine the constant term in the asymptotic expansion, for $t \downarrow 0$, of the supertrace of

$$e^{-t\,\mathrm{Cas}}(4\pi t)^{-\frac{1}{2}\dim P} \int_G e^{-d(p,\,g\,p)^2/4t}\, j(p,\, g\,p)^{-1/2} \cdot \sum_{k=0}^{\infty} t^k\, A_k(p,\, g\,p) \circ \rho(g)\, \mathrm{d}g. \tag{9.26}$$

Here $j = j_P$ is the determinant as defined in (8.44), but now for the Riemannian structure in $P$. The leading term $A_0(p,\, g\,p)$ is equal to the parallel transport in the bundle $\pi^* L$ along the geodesic from $g\,p$ to $p$, defined by the connection (9.21). It is an important remark that, due to (9.22), all the terms $A_k(p,\, g\,p)$ are of the form $1 \otimes T_k(p,\, g\,p)$, where $T_k(p,\, g\,p) \in \mathrm{End}\,(L_x)$, $x = \pi(p)$. The action on $E$ has been transferred to the factor $\rho(g)$, which acts on the fixed vector space E.

In the integral (9.26), one only gets contributions which are not rapidly decreasing, which come from $g$ in a shrinking neighborhood of $g = 1$, so it becomes natural to make the substitution of variables $g = \exp X$, where $X$ varies in a small neighborhood of 0 in $\mathfrak{g}$. The following lemma is a preparation for the computation in Lemma 9.6, of the tangent map of the exponential mapping at $X_{P,p} \in \mathrm{T}_p\, P$.

**Lemma 9.5** *Let $P$ be a Riemannian manifold and let $v$ be a vector field in $P$, such that the solution curves of $v$ are geodesics,* and *that the flow $e^{t\,v}$ of $v$ consists of isometries. For each $p \in P$, let $B(p)$ denote the linear endomorphism*

$$B(p) : u \mapsto -2\,(\nabla_u v)(p).$$

*Then*

$$\mathrm{T}_{v(p)}\mathrm{exp}_p = \mathrm{T}_p e^v \circ \frac{e^{B(p)} - 1}{B(p)}. \tag{9.27}$$

**Proof** Let $\tilde{v}$ denote the vector field in $\mathrm{T}\, P$, which has flow equal to $t \mapsto \mathrm{T}\, \mathrm{e}^{t\, v}$. Let $w$ denote the geodesic vector field in $\mathrm{T}\, P$, as in (8.23). The condition that the $v$-flow consists of isometries means that $\tilde{v}$ commutes with $w$. The condition that the $v$-solution curves $t \mapsto p(t)$ are geodesics means that the curves $t \mapsto (p(t),\, p'(t))$ in $\mathrm{T}\, P$, which by definition are solution curves of $\tilde{v}$, are also solution curves of $w$. In other words, the vector field $w - \tilde{v}$ is identically equal to zero on the submanifold

$$v(P) = \{(p,\, v(p)) \mid p \in P\}$$

of $\mathrm{T}\, P$, the graph of the vector field $v$.

If $v(p) = 0$, then the $\mathrm{T}_p\, \mathrm{e}^{t\, v}$ form a one-parameter group of rotations in $\mathrm{T}_p\, P$, so the solution curves of $v$ near $p$ can only be geodesics if $(\nabla v)(p) = 0$. We therefore may assume that $v(p) \neq 0$, and choose local coordinates $a$ near $p$ in which the vector field $v$ is constant. In such local coordinates the $v$-flow for $t = 1$ is a translation, so its tangent map from $\mathrm{T}_p\, P = \mathbf{R}^d$ to $\mathrm{T}_{\exp_p v(p)}\, P = \mathbf{R}^d$ is equal to the identity. Here $d = \dim P$.

In the corresponding local coordinates $(a,\, b)$ in $\mathrm{T}\, P$, the linearization of $w - \tilde{v}$ is equal to the right hand side of the system of ordinary differential equations

$$\frac{da}{dt} = b, \quad \frac{db}{dt} = A\, a + B\, b,$$

for some $d \times d$-matrices $A$ and $B$. And the tangent space to $v(P)$ now corresponds to $b = 0$. The condition, that $w - \tilde{v} = 0$ on $v(P)$ implies that $A = 0$. So we get that $b - B\, a = c$ is a constant and that

$$\frac{da}{dt} = B\, a + c.$$

The method of variations of constants of Lagrange yields the solution

$$a(t) = \mathrm{e}^{t\, B}\, a(0) + \int_0^t \mathrm{e}^{(t-s)\, B}\, (c)\, \mathrm{d}s.$$

For the computation of $\mathrm{T}_{v(p)} \exp_p$, we take $a(0) = 0$, so $c = b(0)$, and $t = 1$, and we get

$$\mathrm{T}_{v(p)} \exp_p = \frac{\mathrm{e}^B - 1}{B}.$$

We shall now prove that, for each $j$ and $k$

$$(B\,\partial_j,\,\partial_k) = -2\left(\nabla_{\partial_j} v,\,\partial_k\right) \quad \text{at } p. \tag{9.28}$$

Here $\partial_j = \partial/\partial p_j$, viewed both as a differential operator, and as a (constant) vector field. This then completes the proof of (9.27).

A computation in local coordinates with (8.23) shows that

$$\partial_j\,(v,\,\partial_k) + v\,(\partial_j,\,\partial_k) + (B\partial_j,\,\partial_k) = \partial_k\,(v, \partial_j)\,.$$

On the other hand, the fact that $v$ is an infinitesimal isometry, and that we had chosen the local coordinates such that the vector field $v$ is constant, together imply that

$$v\,(\partial_j,\,\partial_k) = ([v,\,\partial_j],\,\partial_k) + (\partial_j,\,[v,\,\partial_k]) = 0.$$

Substituting

$$\partial_j\,(v,\,\partial_k) = \left(\nabla_{\partial_j} v,\,\partial_k\right) + \left(v,\,\nabla_{\partial_j}\partial_k\right)$$

and the same with $j$ and $k$ interchanged, and using that

$$\nabla_{\partial_j}\partial_k - \nabla_{\partial_k}\partial_j = [\partial_j,\,\partial_k] = 0,$$

we get

$$(B\,\partial_j,\,\partial_k) = (\nabla_{\partial_k} v,\,\partial_j) - \left(\nabla_{\partial_j} v,\,\partial_k\right) = -2\left(\nabla_{\partial_j} v,\,\partial_k\right).$$

Here we used in the second identity that $v$ commutes with the $\partial_j$, so

$$\left(\nabla_{\partial_j} v,\,\partial_k\right) + (\partial_j,\,\nabla_{\partial_k} v) = (\nabla_v \partial_j,\,\partial_k) + (\partial_j,\,\nabla_v \partial_k) = v\,(\partial_j,\,\partial_k) = 0.$$

This proves (9.28). □

**Lemma 9.6** *In the following statements, we identify the vertical spaces $V_p$ with $\mathfrak{g}$ and write $X$ instead of $X_{P,p}$. With these notations, we have*

a) *The curves $s \mapsto (\exp s X)\, p$, the orbits in $P$ of the one-parameter subgroups of $G$, are geodesics in $P$.*

b) *For each $p \in P$ and $X \in \mathfrak{g}$, $\mathrm{T}_X \exp_p$ maps $H_p \simeq \mathrm{T}_x M$ to $H_{\exp X(p)} \simeq \mathrm{T}_x M$ and $V_p \simeq \mathfrak{g}$ to $V_{\exp X\,(p)} \simeq \mathfrak{g}$, and*

$$\mathrm{T}_X \exp_p|_{H_p} = J_M\,(X \cdot \Omega_p) := \frac{1 - \mathrm{e}^{-X \cdot \Omega_p}}{X \cdot \Omega_p}, \tag{9.29}$$

$$\mathrm{T}_X \exp_p|_{V_p} = J_{\mathfrak{g}}(\mathrm{ad} X) := \frac{1 - \mathrm{e}^{-\mathrm{ad} X}}{\mathrm{ad} X}. \tag{9.30}$$

c) *The parallel transport along the geodesic $s \mapsto \exp(s\,X)(p)$, with $s$ running from 0 to 1, is equal to the linear transformation*

$$A_0(\exp X\,(p),\, p) = \mathrm{e}^{-\frac{1}{2} X \cdot R^1} \tag{9.31}$$

*in $L_x$, $x = \pi(p)$. Here we used the abbreviation*

$$X \cdot R^1 = \sum_k X_k\, R^1_k, \tag{9.32}$$

*in which $X_k = (X,\, e^{\mathfrak{g}}_k)$ is the k-th coordinate of $X \in \mathfrak{g}$ with respect to the orthonormal basis $e^{\mathfrak{g}}_i$, and the $R^1_k$ are as in (9.18).*

d) *The Jacobian of the substitution $g = \exp X$ is equal to*

$$\mathrm{d}\,(\exp X) = \det\left(\frac{1 - \mathrm{e}^{-\mathrm{ad} X}}{\mathrm{ad} X}\right)\, \mathrm{d}X. \tag{9.33}$$

**Proof** a) follows from (9.5) with $Y = X$.

Since $G$ also acts by isometries on $P$, we can apply Lemma 9.5. In view of (9.6) and (9.5), this leads to b)[1].

Note that (9.30) coincides with the familiar formula

$$\frac{d}{d\epsilon} \exp(-X) \circ \exp(X + \epsilon Y)|_{\epsilon=0} = \frac{1 - \mathrm{e}^{-\operatorname{ad} X}}{\operatorname{ad} X} Y$$

for the derivative of the exponential mapping from the Lie algebra to the Lie group; this in turn implies d).

Finally we observe for c) that, for all $s \in \mathbf{R}$,

$$\pi\left(\exp(s\,X)\,p\right) = x,$$

so the differential equation $\tilde{\nabla}_X \lambda = 0$, cf. (9.21), is a linear differential equation with constant coefficients. The operator which assigns to the initial values at $s = 0$ the solution at $s = 1$, is given by (9.31). □

We have now made all the preparations for the main result of this chapter:

**Theorem 9.1** *With the assumptions as in Lemma 9.3, we get that the integral kernel of* $\mathrm{e}^{-tQ}$ *along the diagonal* $x = y$ *has the expansion*

$$\mathrm{e}^{-t\,\mathrm{Cas}}\,(4\pi t)^{-\frac{1}{2}\dim P} \int_{\mathfrak{g}} \mathrm{e}^{-\|X\|^2/4t} \sum_{k=0}^{\infty} t^k\, B_k(p,\,X) \circ \mathrm{e}^{\rho'(X)}\,\mathrm{d}X. \tag{9.34}$$

*Here* $p \in P$ *is such that* $\pi(p) = x$, *and the coefficients* $B_k$ *satisfy*

$$B_k(p,\,X) \in 1 \otimes \mathrm{End}\,(L_x)\,, \quad x = \pi(p), \tag{9.35}$$

*and depend smoothly on* $p$ *and* $X$. *Furthermore,*

$$B_0(p,\,X) = j_{\mathfrak{g}}(\mathrm{ad}X)^{1/2}\, j_M(-X \cdot \Omega_p)^{-1/2}\left(1 \otimes \mathrm{e}^{\frac{1}{2}X \cdot R^1}\right), \tag{9.36}$$

[1] In the formula in [9, Th. 5.4], corresponding to (9.30), the sign of $a$ seems to be wrong, because in [9, (5.2)] a left multiplication by $\exp(-a)$ has been used, whereas the action on $P$ in [9, Th. 5.4] is from the right. For the computation of $j_{\mathfrak{g}}(\operatorname{ad} X)$ in Theorem 9.1 below, this has no effect, because the adjoint of $J_{\mathfrak{g}}(\operatorname{ad} X)$ is equal to $J_{\mathfrak{g}}(-\operatorname{ad} X)$, and these have the same determinants.

*in which*

$$j_{\mathfrak{g}}(\mathrm{ad}X) := \det J_{\mathfrak{g}}(\mathrm{ad}X) = \det \frac{1 - \mathrm{e}^{-\mathrm{ad}X}}{\mathrm{ad}X}, \tag{9.37}$$

$$j_M(-X \cdot \Omega_p) := \det J_M(-X \cdot \Omega_p) = \det \frac{\mathrm{e}^{X \cdot \Omega_p} - 1}{X \cdot \Omega_p}. \tag{9.38}$$

*(For the notations, see also (9.4) and (9.32).)*

*The integration in (9.34) is only performed over a small neighborhood of the origin in* $\mathfrak{g}$*; another cutoff yields the same expansion.*

**Proof** With the substitution $g = \exp X$, we get

$$d(p, gp)^2 = \|X\|^2, \quad \text{and} \quad \rho(g) = \mathrm{e}^{\rho'(X)}.$$

Also,

$$j\,(p, gp) = j\left(g^{-1}p, p\right) = j\,(\exp -X_P\,p, p)\,.$$

For the parallel transport, we have:

$$A_0(p, gp) = A_0(gp, p)^{-1} = \left(\mathrm{e}^{-\frac{1}{2}X \cdot R^1}\right)^{-1}.$$

Using Lemma 9.6, we see that (9.26) gets the form as described in the proposition. □

In Chapter 11, we will determine the constant term, for $t \downarrow 0$, of the supertrace of (9.34), in the case that $Q = D^2$ and $D$ is the spin-c Dirac operator. This will yield the formula (11.17) for the integrand in the integral formula (7.11) for the index of $D^+$. For this, one may skip Chapter 10, which is a preparation for the local Lefschetz formula in the presence of an automorphism $\gamma$.

# Chapter 10

# The Automorphism

The goal here is Theorem 10.1 below, which is the modification of Theorem 9.1 in the presence of an automorphism $\gamma$. The only contributions to the asymptotic expansion come from neighborhoods of fixed points in $M$ of $\gamma$, which in the principal $G$-bundle $P$ correspond to fixed points modulo the action of $G$. The extra ingredient needed here is the asymptotic estimate in Lemma 10.1 of the geodesic distance between two points in $P$ which are close to the same $G$-orbit.

## 10.1 Assumptions

Let $\gamma = \gamma_M$ be a transformation in $M$ which preserves all the geometric data used to define the spin-c Dirac operator $D$. That is, for each $x \in M$, we have

$$\mathrm{T}_x\,\gamma \circ J_x = J_{\gamma(x)} \circ \mathrm{T}_x\,\gamma. \tag{10.1}$$

Furthermore, $\gamma$ is an isometry in $M$ with respect to the Riemannian structure $\beta$. And finally, we assume that there is a bundle automorphism $\gamma_L$ of $L$, which preserves the Hermitian structure and the connection in $L$, and which *covers* $\gamma_M$, in the sense that $\pi_L \circ \gamma_L = \gamma_M \circ \pi_L$, if $\pi_L$ denotes the projection from $L$ onto $M$.

**Remark** According to a theorem of van Dantzig and van der Waerden [25], the action of the isometry group $I$ of a connected Riemannian manifold $M$ is *proper*. That is,

$$(\phi,\, x) \mapsto (\phi(x),\, x)$$

is a proper mapping from $I \times M$ to $M \times M$. We may assume that $\gamma$ has a fixed point $x$, so $\gamma \in I_x$ if $I_x$ denotes the set of $\phi \in I$ such that $\phi(x) = x$. It follows now from the properness of the action of $I$ that the group $I_x$ is compact.

In the applications, it may occur that $\gamma$ is an automorphism of the almost complex manifold $(M,\ J)$, covered by an automorphism $\gamma_L$ of $L$, and that $\gamma_L$ belongs to a compact group $K$ of transformations. By averaging an arbitrary Riemannian structure in $M$ and Hermitian structure in $L$ over $K$, one arrives at a Riemannian structure in $M$ and Hermitian structure in $L$, which are invariant under $\gamma$ and $\gamma_L$, respectively.

From a given $\gamma$-invariant Riemannian structure $\beta$ en $\gamma$-invariant nondegenerate two-form $\sigma$, one can also get a $\gamma$-invariant almost complex structure $J$ by taking

$$J_x = \sigma_x^{-1} \circ \beta_x \circ {P_x}^{-1},$$

in which $P_x$ is the positive square root of the positive definite symmetric transformation $-(A_x)^2$. (See (15.41) for more details.) In this way, the theory is applicable to transformations $\gamma$, which belong to a compact group of automorphisms of a symplectic manifold $(M,\ \sigma)$.

The condition that $\gamma$ is an isometry of $(M,\ \beta)$, in combination with (10.1), yields that the mapping

$$\mathrm{T}\gamma : e \mapsto \mathrm{T}\gamma \circ e \tag{10.2}$$

defines a bundle automorphism $\mathrm{T}\gamma$ of UF$\,M$ which covers $\gamma$. We have a similar bundle automorphism of SOF$\,M$, also denoted by $\mathrm{T}\gamma$. The embedding $e \mapsto f(e)$ of UF$\,M$ into SOF$\,M$ intertwines the two $\mathrm{T}\gamma$'s, so there is not much danger of confusion.

The transformation $\mathrm{T}\gamma$ commutes with the action $e \mapsto e \circ A^{-1}$ of $A \in \mathrm{U}(n)$, so it induces a transformation $\gamma_P$ in the principal $\mathrm{Spin}^{\mathrm{c}}$-bundle

$$P = \mathrm{Spin}^{\mathrm{c}}\,\mathrm{F}\,M,$$

which commutes with the action of $G = \mathrm{Spin}^{\mathrm{c}}(2n)$ in $P$. Also, because $\gamma$ is an isometry of $(M, \beta)$, the transformation $\mathrm{T}\gamma$ in $\mathrm{SOF}\,M$ leaves the Levi-Civita connection in $\mathrm{SOF}\,M$ invariant. And because $\mathrm{T}\gamma$ commutes with $J$, it leaves the Hermitian structure and the connection in $K^*$ invariant. Since the connection in $P$ was defined as the pullback of the connection in $\Delta^*\,(\mathrm{SOF}\,M \times \mathrm{U}(K^*))$ by means of the covering $\gamma$ in (5.6) (sorry for the clash of notations), it follows that $\gamma_P$ leaves the connection in $P$ invariant. Finally, we observe that the induced action of $\gamma$ on $E$, which we could also denote by $\mathrm{T}\gamma$, coincides with the one on (5.7), obtained by letting $\gamma_P$ act on the first factor.

In the general setting of the previous chapter, we assume now that the Riemannian manifold $(M, \beta)$ is oriented, and that $\gamma$ is an orientation preserving isometry of $(M, \beta)$. Furthermore, it is covered by a bundle automorphism $\gamma_P$ of $P$ which commutes with the $G$-action in $P$ and preserves the connection in $P$. And finally, $\gamma$ is also covered by a bundle automorphism $\gamma_L$ of $L$, which preserves the Hermitian structure and connection of $L$.

These data lead to the definition of the operator $\gamma_{\mathrm{op}}$, acting on sections $\omega$ of $E \otimes L$, by means of

$$\left(\gamma_{\mathrm{op}}(\omega \otimes \lambda)\right)(p) = \omega\left(\gamma_P^{-1}(p)\right) \otimes \gamma_{L,\,y}\,\lambda_y, \quad p \in P,\ y = \gamma^{-1} \circ \pi(p) \in M. \tag{10.3}$$

It is clear from all the invariance properties of $\gamma$ mentioned above that $\gamma_{\mathrm{op}}$ commutes with $Q$, Cas, and $\tilde{Q}$. So, in the same way as we arrived at (9.24), we now get that the integral kernel of

$$\gamma_{\mathrm{op}} \circ \mathrm{e}^{-tQ} = \mathrm{e}^{-tQ} \circ \gamma_{\mathrm{op}} \tag{10.4}$$

is equal to

$$\mathrm{e}^{-t\,\mathrm{Cas}} \int_G K^P\left(t,\, p,\, g\,\gamma_P(q)\right) \circ \rho(g) \circ \left(1 \otimes \gamma_{L,\,\pi(q)}\right)\, dg. \tag{10.5}$$

For the study of (7.6), we may restrict to $p = q$. Substituting the asymptotic expansion (8.48), we see that we only get contributions which are not decreasing as $t \downarrow 0$ , if the distance between $p$ and $g\,\gamma_P(p)$ is of order $t^{1/2}$. Since $\gamma_P$ covers $\gamma$, this implies that the distance between $x = \pi(p)$ and $\gamma(x)$ also is of order $t^{1/2}$, or that $x$ is at a distance of order $t^{1/2}$ to a fixed point $x_0$ of $\gamma$ in $M$.

In a neighborhood of a fixed point $x_0$, and in normal coordinates, the isometry $\gamma = \gamma_M$ of $(M,\,\beta)$ has indeed a very simple description. It maps geodesics to geodesics, whereby the velocity vectors are transformed by means of $\mathrm{T}\,\gamma$. (In other words, the transformation $\mathrm{T}\,\gamma$ in $\mathrm{T}\,M$ commutes with the geodesic flow in $\mathrm{T}\,M$.) In particular we have

$$\gamma\left(\exp_{x_0} v\right) = \exp_{x_0}\left(\mathrm{T}_{x_0}\,\gamma(v)\right), \quad v \in \mathrm{T}_{x_0}\,M. \tag{10.6}$$

That is, the exponential mapping, centered at the fixed point $x_0$, intertwines the linear transformation $\mathrm{T}_{x_0}\,\gamma$ in the tangent space $\mathrm{T}_{x_0}\,M$, with the transformation $\gamma$ in a neighborhood $U$ of $x_0$ in $M$. Note that $\mathrm{T}_{x_0}\,\gamma$ leaves the inner product $\beta_{x_0}$ in $\mathrm{T}_{x_0}\,M$ invariant. And that the assumption that $\gamma$ is an orientation preserving isometry implies that $\mathrm{T}_{x_0}\,\gamma$ is an orientation preserving orthogonal linear transformation, a rotation, in $\mathrm{T}_{x_0}\,M$.

The fixed point set of $\mathrm{T}_{x_0}\,\gamma$ in $\mathrm{T}_{x_0}\,M$ is a linear subspace $\mathrm{T}_{x_0}^{\gamma}\,M$ of $\mathrm{T}_{x_0}\,M$. The orthogonal complement $N_{x_0}$ of $\mathrm{T}_{x_0}^{\gamma}\,M$ in $\mathrm{T}_{x_0}\,M$ is invariant under $\mathrm{T}_{x_0}\,\gamma$, with 0 as the only fixed point. If we write

$$\gamma_N := \mathrm{T}_{x_0}\,\gamma|_{N_{x_0}}, \tag{10.7}$$

this means that

$$\gamma_N - 1 : N_{x_0} \to N_{x_0} \quad \text{is invertible.} \tag{10.8}$$

In view of the decomposition of the rotation $\mathrm{T}_{x_0}\,\gamma$ into planar rotations, this implies that the dimension of $N_{x_0}$ is even, say equal to $2l$.

These facts imply that the fixed point set $U^{\gamma} = M^{\gamma} \cap U$ of $\gamma$ in $U$ is a smooth submanifold of $M$, of even codimension equal to $2l$. Globally, this implies that each connected component $F$ of $M^\gamma$ is a closed, smooth submanifold of $M$. We also observe that the eigenvalues of $\mathrm{T}_{x_0}\gamma$ do not change, as $x_0$ varies in $F$. The only warning is that in general, the normal action of $\gamma$ will be different along different connected components $F$ of $M^\gamma$, and even the codimensions $l = l_F$ of the $F$ can vary.

Using a partition of unity, with smooth functions with support in normal coordinate neighborhoods as above, the study of the integral (7.6) is reduced to an integral over $U$, where the integrand has been multiplied by a test function $\psi(x)$, which is a smooth real-valued function with compact support in $U$. We will perform the integration by first integrating over $N_{x_0}$ and then over $U^\gamma$. Actually, we will not need to perform the integration over $U^\gamma$, because it will turn out that the integral over $N_{x_0}$ already has an asymptotic expansion in powers of $t$, with an explicit expression for the leading term. Moreover, the remainder term in the asymptotic expansion is of order $t^k$ for large $k$, locally uniformly when $x_0 \in F$ varies. Also, the asymptotic expansion can arbitrarily often be differentiated termwise with respect to both $x_0$ and to $t$.

## 10.2 An Estimate for Geodesics in $P$

In order to prove the announced results on the integration over $N_{x_0}$, we first need a quite precise estimate for the distance between $p$ and $g\,\gamma_P(p)$, cf. Lemma 10.1 below.

Recall that we need only to consider points $p$, which are close to a point $p_0 \in P$, where $x_0 = \pi(p_0)$ is a fixed point of $\gamma$ in $M$. But, in general, this does not imply that $p_0$ is a fixed point for $\gamma_P$ in $P$. Indeed, in our situation, $\gamma_P$ is the transformation in $P$ which is induced by the transformation $\mathrm{T}\,\gamma$ in $\mathrm{T}\,M$, which leaves $\mathrm{T}_{x_0} M$ invariant but has no nonzero fixed points in the linear subspace $N_{x_0}$. We see that unless $\gamma = 1$, $\mathrm{T}\,\gamma$ acts without fixed points on $\mathrm{UF}\,M$, and hence also $\gamma_P$ has no fixed points in $P$.

However, the fact that $\gamma(x_0) = x_0$ is equivalent to the condition that $\gamma_P(p_0) \in G\,p$, or to the existence of a (uniquely determined) $g_0 \in G$, such that $g_0\,\gamma_P(p_0) = p_0$. We will therefore write

$$g \circ \gamma_P = h \circ \delta, \quad \text{with} \quad h = g\,{g_0}^{-1}, \quad \delta = g_0 \circ \gamma_P. \tag{10.9}$$

Then $\delta$ is a transformation in $P$, still covering $\gamma$, such that $\delta(p_0) = p_0$. We will subsequently perform the substitution of variables $g = h\,g_0$ in the integral (10.5). Since now we can restrict to $h$ close to the identity element in $G$, we subsequently will substitute $h = \exp X$, with $X$ in a small neighborhood of 0 in $\mathfrak{g}$, in order to arrive (after the integration over $N_{x_0}$) at an integral which is similar to (9.34).

In the integration over $N_{x_0}$, we have an exponential factor, with in the exponent $-1/4t$ times the square of the distance from $p$ to $(\exp X_P)\,(q)$, with $p$ and

$$q = \delta(p) = g_0\,\gamma_P(p) \tag{10.10}$$

close to each other. The integration over $N_{x_0}$ means that we take $x = \exp_{x_0} v$, where $v$ varies in a small neighborhood of the origin in $N_{x_0}$, and then choose $p \in P$ such that $\pi(p) = x$. A natural choice is to take

$$p(v) = \exp_{p_0}(v_{\text{hor}})\,, \tag{10.11}$$

in which $v_{\text{hor}}$ is the horizontal lift of $v$ in $\mathrm{T}_{p_0}\,P$, the vector in $H_{p_0}$ which projects to $v$. Note that the choice (9.3) of the Riemannian structure in $P$ gives that $v_{\text{hor}}$ is orthogonal to $X_{P,\,p_0}$. The curve $t \mapsto p(t\,v)$ is a geodesic, starting at $p_0$, with horizontal tangent vector $v$. Since both $\gamma_P$ and $g_0$ are isometries which preserve the connection, $t \mapsto \delta\,(p(t\,v))$ is also a geodesic, starting at $p_0$, this time with the horizontal tangent vector

$$u = \mathrm{T}_{x_0}\,\gamma(v), \tag{10.12}$$

if we identify all horizontal vectors over $x_0$ with their projections in $\mathrm{T}_{x_0}\,M$. The following lemma serves to get a sufficiently accurate estimate for the geodesic distance between $p(t\,v)$ and $(\exp X) \circ \delta\,(p(t\,v))$.

**Lemma 10.1** *For $X \in \mathfrak{g}$, $p \in P$ and $v, w \in H_p$, we have the locally uniform estimate*

$$d\left(\exp_p(t\,v),\, \exp X\, \exp_p(t\,w)\right)^2 = \|X\|^2$$
$$+t^2\left((v,\, J^{-1}v) - 2\,(v,\, J^{-1}w) + (w,\, J^{-1}w)\right) + \mathcal{O}\left(t^3\right), \quad (10.13)$$

*as $t \to 0$. Here $J = J_M\,(X \cdot \Omega_p)$, as in (9.29).*

**Proof** [1] Write

$$p(t) = \exp_p\,(t\,v)\,. \quad (10.14)$$

It will be convenient to replace $t\,v$ and $t\,w$ by $t_1\,v$ and $t_2\,w$, respectively, with indepedently moving $t_1$ and $t_2$. Let $u(t_1,\, t_2)$ be the element in $\mathrm{T}_{p(t_1)}\,P$, close to $X = X_{P,\,p(t_1)}$, such that

$$\exp_{p(t_1)}\, u(t_1,\, t_2) = q(t_2) := \exp X_P\, \exp\,(t_2\,w)\,. \quad (10.15)$$

This gives that the left hand side in (10.13) is equal to the square of the length of $u(t_1,\, t_2)$. Note that $p(0) = p$, $p'(0) = v$, and $u(0,\, 0) = X$.

In the sequel, we will write $\partial_i = \partial/\partial t_i$. Using (9.29), we get

$$v + J\,\partial_1 u(0,\, 0) = 0, \quad J\,\partial_2 u(0,\, 0) = w. \quad (10.16)$$

Since $J$ leaves $H_p$ invariant and $u(0,\, 0) = X$ is orthogonal to $H_p$, we get that

$$(u(t),\, u(t)) = \|X\|^2 + \mathcal{O}\left(\|t\|^2\right).$$

as $t \downarrow 0$.

Now consider the map $\epsilon$ from an open neighborhood $T$ of $\{0\} \times \{0\} \times [0,\, 1]$ in $\mathbf{R}^3$ into $P$, defined by

$$\epsilon(t_1,\, t_2,\, t_3) = \exp_{p(t_1)}\,(t_3\, u(t_1,\, t_2))\,. \quad (10.17)$$

[1] We follow the clever proof of [9, Prop. 6.17].

We will view $\partial_i$, $i = 1, 2, 3$, depending on the context, either as differential operators, or as vector fields in $T$.

We will also make a somewhat serious use of the *pullback bundle* $\epsilon^* \mathrm{T} P$ of $\mathrm{T} P$. It is defined as the set of $(t, w) \in T \times \mathrm{T} P$, such that $\epsilon(t) = \pi(w)$ if $\pi$ denotes the projection from $\mathrm{T} P$ onto $P$. The projection to the first component defines a projection from $\epsilon^* \mathrm{T} P$ onto $T$, which turns $\epsilon^* \mathrm{T} P$ into a vector bundle over $T$; the fiber over $t \in T$ is identified with $\mathrm{T}_{\epsilon(t)} P$. Sections of $\epsilon^* \mathrm{T} P$ are mappings $s : T \to \mathrm{T} P$, with the property that $\pi \circ s = \epsilon$. These are sometimes also called "vector fields in $P$, lying over $\epsilon$". Examples of such sections are the partial derivatives $\partial_i \epsilon$.

For such a section $s$, the covariant derivative at $t \in T$, in the direction of a vector $\tau \in \mathrm{T}_t T$, is defined by

$$(\nabla_\tau s)(t) = \mathrm{T}_t\, s\,(\tau) - (\mathrm{T}_t\, \epsilon\,(\tau))_{\mathrm{hor},\, s(t)} \,.$$

These definitions can actually be given for an arbitrary connection in an arbitrary bundle over $P$, but we will apply this here only for the case of the Levi-Civita connection in $\mathrm{T} P$. The point of the whole construction is that all formulas for connections (such as concerning torsion and curvature) hold as if $T$ were a submanifold of $P$ and smooth sections of $\epsilon^* \mathrm{T} P$ were restrictions to $T$ of smooth vector fields in $P$.

As a function of $t_3$, the right hand side in (10.17) is a geodesic, so

$$\nabla_{\partial_3} \partial_3 \epsilon = 0.$$

This yields

$$\begin{aligned} {\partial_3}^2\,(\partial_1\epsilon,\, \partial_3\epsilon) &= (\nabla_{\partial_3} \nabla_{\partial_3} \partial_1\epsilon,\, \partial_3\epsilon) \\ &= (\nabla_{\partial_3} \nabla_{\partial_1} \partial_3\epsilon,\, \partial_3\epsilon) = (R\,(\partial_3\epsilon,\, \partial_1\epsilon)\, \partial_3\epsilon,\, \partial_3\epsilon) = 0. \end{aligned}$$

Here we used in the first equation that the Riemannian structure is covariantly constant, in the second equation that the connection is torsion-free, in the third equation that the vector fields $\partial_3$ and $\partial_1$ commute, and in the last equation that the curvature is anti-symmetric with respect to the inner product. It follows that $(\partial_1\epsilon,\, \partial_3\epsilon)$ is an affine function of $t_3$. Since

$$\epsilon(t_1,\, t_2,\, 1) \equiv q(t_2),$$

we get $\partial_1\epsilon = 0$ when $s = 1$. On the other hand, if $t_3 = 0$, then $\partial_1\epsilon = p'(t_1)$ and $\partial_3\epsilon = w(t_1,\, t_2)$, so

$$(\partial_1\epsilon,\, \partial_3\epsilon) = (1 - t_3)\; (p'(t_1),\, u(t_1,\, t_2))\,.$$

From this, we obtain

$$\begin{aligned} &\partial_1\,(u,\, u) = \partial_1\,(\partial_3\epsilon,\, \partial_3\epsilon) = 2\,(\nabla_{\partial_1}\partial_3\epsilon,\, \partial_3\epsilon) \\ &= 2\,(\nabla_{\partial_3}\partial_1\epsilon,\, \partial_3\epsilon) = 2\partial_3\,(\partial_1\epsilon,\, \partial_3\epsilon) = -2\,(p'(t_1),\, u(t_1,\, t_2))\,. \end{aligned}$$

Using that $\nabla_{\partial_1}p' = 0$, because $t_1 \mapsto p(t_1)$ is a geodesic, we get, in combination with (10.16), that

$$\begin{aligned} \partial_1{}^2\,(u,\, u)|_{t_1=t_2=0} &= 2\,\left(v,\, J^{-1}\,v\right), \\ \partial_2\,\partial_1(u,\, u)|_{t_1=t_2=0} &= -2\,\left(v,\, J^{-1}\,w\right). \end{aligned}$$

On the other hand, the geodesic distance between $p_1$ and $\exp X_P\,(p_2)$ is the same as the geodesic distance between $\exp(-X_P)\,(p_1)$ and $p_2$. Since the adjoint of $-X \cdot \Omega_p$ is equal to $X \cdot \Omega_p$, we therefore get that

$$\partial_2{}^2\,(u,\, u)|_{t_1=t_2=0} = 2\,\left(w,\, J\,(-X \cdot \Omega_p)^{-1}\,w\right) = 2\,\left(J\,(X \cdot \Omega_p)^{-1}\,w,\, w\right).$$

This completes the proof of (10.13). □

## 10.3 The Expansion

We are now ready for the proof of the following result.

**Theorem 10.1** *Let $F$ be a connected component of the fixed point set $M^\gamma$ of $\gamma$ in $M$; $F$ is a smooth submanifold of $M$ of codimension $2l_F$ in $M$. Let $x \in F$, and let $\psi$ be a smooth real-valued function, with compact support, contained in a small normal coordinate neighborhood $U$ around $x$. Let $N_x$ denote the orthogonal complement of $\mathrm{T}_xF$ in $\mathrm{T}_xM$. Then the integral over the diagonal in $N_x \times N_x$, of the integral kernel of (10.4), is asymptotically of the form*

$$\mathrm{e}^{-t\,\mathrm{Cas}}\,(4\pi t)^{l_F-\frac{1}{2}\dim P}\int_{\mathfrak{g}}\mathrm{e}^{-\|X\|^2/4t}\sum_{k=0}^{\infty}t^k\,C_k(p,\,X)\circ\mathrm{e}^{\rho'(X)}\circ\rho(g_0)\,\mathrm{d}X. \tag{10.18}$$

*Each of the coefficients $C_k$ satisfies*

$$C_k(p,\,X)\in 1\otimes\mathrm{End}\,(L_x)\,,\quad x=\pi(p), \tag{10.19}$$

*and depends smoothly on $p$ and $X$ and finitely many derivatives of $\psi$ at $x$. Furthermore, if $\psi(x)=1$,*

$$\begin{aligned} C_0(p,\,X) &= j_{\mathfrak{g}}(\mathrm{ad}X)^{1/2}\,j_F\,(-X\cdot\Omega_p)^{-1/2}\left(1\otimes\mathrm{e}^{\frac{1}{2}X\cdot R^1}\,\gamma_{L,x}\right)\\ &\quad\cdot\det\left(1-\gamma_N{}^{-1}\,\mathrm{e}^{-X\cdot\Omega_p}\right)^{-1/2}\det\left(1-\gamma_N\right)^{-1/2}. \end{aligned} \tag{10.20}$$

*Here $j_{\mathfrak{g}}$ is as in (9.37). The space $\mathrm{T}_xF$ is invariant under the mapping $J_M\,(-X\cdot\Omega_p)$, defined in (9.29), and $j_F(-X\cdot\Omega_p)$ denotes the determinant of the restriction $J_F\,(-X\cdot\Omega_p)$ of $J_M\,(-X\cdot\Omega_p)$ to $\mathrm{T}_xF$. Finally, $\gamma_N$ is restriction of $\mathrm{T}_x\gamma$ to $N_x$.*

*The integration in (10.18) is performed only over a small neighborhood of the origin in $\mathfrak{g}$; another cutoff yields the same expansion.*

**Proof** We resume the discussion in the paragraphs preceding Lemma 10.1. With $p(v)$ as in (10.11), $v\in H_x$, we get from (10.12) and (10.13) that

$$\phi(v):=d\,(p(v),\,\exp X_P\,\delta(p(v)))^2-\|X\|^2$$

satisfies $\phi(v) = q(v) + \mathcal{O}(\|v\|^3)$ as $v \to 0$, where $q(v)$ is the quadratic form

$$q(v) = \left(v,\, J^{-1} v\right) - 2\left(v,\, J^{-1} \gamma_N v\right) + \left(\gamma_N v,\, J^{-1} \gamma_N v\right).$$

We will use the fact that $\gamma_N$ and $J$ commute, and that $\gamma_N$ is an isometry, in order to simplify $q(v)$ to

$$q(v) = 2\left(v,\, J^{-1}\,(1 - \gamma_N)\, v\right).$$

Next, we will write it in the form $q(v) = (Q\, v,\, v)$ for the symmetric operator

$$Q = Q(X) := J^{-1}\,(1 - \gamma_N) + (1 - \gamma_N{}^*)\left(J^{-1}\right)^*. \tag{10.21}$$

For $X = 0$ we have $J = 1$, and because

$$1 - \gamma_N{}^* = -\gamma_N{}^*\,(1 - \gamma_N), \tag{10.22}$$

we see that

$$Q(0) = (1 - \gamma_N)^*\,(1 - \gamma_N)$$

is positive definite in $N_x$. It follows that, for all $X$ in a neighborhood of 0 in $\mathfrak{g}$, $Q(X)$ is still positive definite.

By the Morse lemma with parameters, we can therefore find a regular substitution of variables $v = v(w)$, smoothly depending on $p$ and $X$, such that $v(0) = 0$ and $\phi(v(w)) = \|w\|^2$. If we write $A = \frac{dv(w)}{dw}|_{w=0}$, then we get $A^*\, Q\, A = 1$; so for the Jacobian at $w = 0$ we get

$$\det \tfrac{dv(w)}{dw}|_{w=0} = \det Q(X)^{-1/2}. \tag{10.23}$$

The integration of $K^P$ in (9.24) over $v$ therefore amounts to an integration of the form

$$I = (4\pi t)^{-l_F} \int_{N_x} \mathrm{e}^{-\|w\|^2/4t}\, a(w,\, x,\, X,\, t)\, dw,$$

where we borrowed the appropriate power of $4\pi t$ from the power in front of the asymptotic expansion for $K^P$. (Recall that $\dim(N_x) = 2l_F$.) The amplitude $a$ has an asymptotic expansion of the form

$$a(w,\, x,\, X,\, t) \sim \sum_{k=0}^{\infty} t^k\, a_k(w,\, x,\, X)$$

as $t \downarrow 0$, which is smooth and locally uniform in the parameters. The substitution $w = t^{1/2}\, z$ now yields an expansion of the form

$$I \sim \sum_{k=0}^{\infty} t^k\, I_k(x,\, X), \quad t \downarrow 0$$

for the integral, which is locally uniform and can be differentiated termwise with respect to all the parameters. Furthermore, $I_0(x,\, X) = a_0(0,\, x,\, X)$, whereas the $I_k$ for $k > 0$ are expressions in corresponding higher order Taylor expansions at $w = 0$, of the $a_l(w,\, x,\, X)$ with $l \le k$.

For the identification of (10.20), we observe that[2]:

$$J^* = J_M\,(-X \cdot \Omega_p) = \frac{\mathrm{e}^{X \cdot \Omega_p} - 1}{X \cdot \Omega_p} = \mathrm{e}^{X \cdot \Omega_p}\; J_M\,(X \cdot \Omega_p)\,. \tag{10.24}$$

Using (10.22), and the facts that $X \cdot \Omega_p$ commutes with $\gamma_N$, and that ${\gamma_N}^* = {\gamma_N}^{-1}$, we can rewrite (10.21) as

$$Q(X) = J^{-1}\left(1 - {\gamma_N}^{-1}\, \mathrm{e}^{-X \cdot \Omega_p}\right)(1 - \gamma_N) \quad \text{on } N. \tag{10.25}$$

The factor $J^{-1}$ in (10.25) leads to a factor

$$(\det\nolimits_{N_x} J(X \cdot \Omega_p))^{1/2} = (\det\nolimits_{N_x} J(-X \cdot \Omega_p))^{1/2}$$

in $\det Q(X)^{-1/2}$, which combined with the factor $(\det J(-X \cdot \Omega_p))^{-1/2}$ which we had in the expansion of $K^P$ leads to the factor $j_F\,(-X \cdot \Omega_p)^{-1/2}$ in (10.20).

---

[2] It seems that in the formula for $A^*$ in [9, p. 205], a wrong sign of $a$ has crept in.

We finally observe that since $\gamma_N$ is a rotation without nonzero fixed points, the determinant of $1 - \gamma_N$ is positive. In order to prove this, we use the decomposition into planar rotations. For a planar rotation $R$ over the angle $\alpha$, the determinant of $1 - R$ is equal to

$$(1 - \cos\alpha)^2 + \sin^2\alpha = 2\,(1 - \cos\alpha) > 0$$

if $\alpha$ is not equal to an integral multiple of $2\pi$. It follows that, for small $X$, also the determinant of $1 - {\gamma_N}^{-1}\,\mathrm{e}^{-X\cdot\Omega_p}$ is positive. And because the determinant of $Q(X)$ was positive, the determinant of the restriction to $N$ of $J^{-1}$ is positive as well. This allows us to take (positive) square roots of each of these determinants, completing the proof of (10.20). □

In the special case of the spin-c Dirac operator, the linear transformation $\mathrm{T}_{x_0}\,\gamma$ commutes with the almost complex structure $J_{x_0}$ in $\mathrm{T}_{x_0}\,M$, cf. (10.1). (Sorry for the clash of notation with $J = J_M\,(X \cdot \Omega_p)$.) That is, $\mathrm{T}_{x_0}\,\gamma$ is a unitary transformation in $\mathrm{T}_{x_0}\,M$. It follows that both $\mathrm{T}_{x_0}\,F$ and $N_{x_0}$ are invariant under $J_{x_0}$. That is, $J$ restricts to an almost complex structure in $F$, and $\gamma_N$ is a complex linear transformation in $N_{x_0}$, with all eigenvalues on the unit circle and not equal to one.

In Chapter 12, we will see that, again for the spin-c Dirac operator, the integral over $N_{x_0}$ of the supertrace converges for $t \downarrow 0$. The limit is an expression in terms of the geometric data at $x_0$, and the test function $\psi$ enters only via multiplication by $\psi(x_0)$. That is, no derivatives of $\psi$ at $x_0$ appear. This means that the function

$$x \mapsto \mathrm{str}_{\mathbf{c}}\,K(t,\,x,\,x)$$

converges, for $t \downarrow 0$, in distribution sense, and the limit is a smooth density in each connected component $F$ of $M^\gamma$.

# Chapter 11

# The Hirzebruch-Riemann-Roch Integrand

## 11.1 Introduction

In this chapter, we will determine the constant term, for $t \downarrow 0$, of the supertrace of (9.34) in the case that $Q = D^2$ and $D$ is the spin-c Dirac operator. This will yield the formula (11.17) in Theorem 11.1 for the integrand, in the integral formula (7.11) for the index of $D^+$.

That is, in (9.34) we now take $\mathrm{E} = \mathrm{E}(n)$, the direct sum over $q$ of the space of $(0,\ q)$-forms on $\mathbf{C}^n$. The group $G$ is equal to $\mathrm{Spin}^{\mathrm{c}}(2n)$, cf. Chapter 4, acting on $\mathrm{E}(n)$ by means of the Clifford multiplication c. Since both $G$ and $\mathfrak{g}$ appear as subsets of the complex Clifford algebra $\mathrm{C}(2n) \otimes \mathbf{C}$, and c is a linear mapping from $\mathrm{C}(2n) \otimes \mathbf{C}$ to $\mathrm{End}\,(\mathrm{E}(n))$ (actually, an isomorphism of algebras), we can write $\rho = \mathrm{c}\,|_G$ and $\rho' = \mathrm{c}\,|_{\mathfrak{g}}$.

The Clifford multiplication by elements of $G$ and $\mathfrak{g}$, respectively, leaves the decomposition of $\mathrm{E}(n)$ into its even part $\mathrm{E}(n)^+$ and its odd part $\mathrm{E}(n)^-$ invariant, so the same is true for the integrand in (9.34). In the integral (7.4), the integrand is the supertrace of the integral kernel along the the diagonal, which is the trace of the restriction to $E^+ \otimes L$ minus the trace of the restriction to $E^- \otimes L$. Therefore we have to take the supertrace of (9.34), which here is the trace of the endomorphism of $\mathrm{E}(n)^+ \otimes L_x$, minus the trace of the endomorphism of $\mathrm{E}(n)^- \otimes L_x$.

One of the keys is formula (3.33), which expresses the supertrace of an endomorphism of $E$ in terms of the volume part of a corresponding element of the exterior algebra. Although the correspondence is not a homomorphism of algebras, it is close enough to get that, for the computation of the leading term in the asymptotic expansion, one may replace the algebra of endomorphisms by the exterior algebra. For the precise statements to this effect, see Lemma 11.2 through 11.6.

The translation of the formulas in terms of characteristic classes will be discussed in Chapter 13, cf. Proposition 13.1. The modifications in the presence of an automorphism $\gamma$ are treated in Chapter 12.

In this chapter and the next, we will use the notation of Chapters 3 and 4, and write $V$, $\mathrm{C}(V)$, $E$, $\mathrm{Spin}(V)$, and $\mathrm{Spin}^{\mathrm{c}}(V)$ instead of $\mathbf{R}^{2n} \simeq \left(\mathbf{R}^{2n}\right)^*$, $\mathrm{C}(2n)$, $\mathrm{E}(n)$, $\mathrm{Spin}(2n)$, and $\mathrm{Spin}^{\mathrm{c}}(2n)$, respectively.

Several of the factors in (9.34) will turn out to be quite innocuous. To begin with, the Casimir element $\mathrm{Cas} \in \mathrm{End}(E)$ actually acts as a scalar multiplication. In order to prove this, we observe that it is equal to the c-image of the element

$$\mathrm{Cas}_{\mathbf{C}} := \sum_k {e_k}^2 \in \mathrm{C}(V) \otimes \mathbf{C}, \tag{11.1}$$

where $e_k = e_k^{\mathfrak{g}}$ denotes an orthonormal basis in $\mathfrak{g} \subset \mathrm{C}(V) \otimes \mathbf{C}$, cf. (9.15). The statement is therefore equivalent to the fact that $\mathrm{Cas}_{\mathbf{C}}$ is a scalar, cf. Lemma 11.1 below. In the proof, we also introduce a special kind of orthonormal basis in $\mathfrak{g}$, which will be used throughout this chapter.

**Lemma 11.1** $\mathrm{Cas}_{\mathbf{C}}$ *is a scalar.*

**Proof** The Lie algebra of $\mathfrak{g} = \mathfrak{spin}^{\mathrm{c}}(V)$ is equal to the direct sum of the Lie algebra $\mathfrak{spin}(V)$ of $\mathrm{Spin}(V)$ and $\mathfrak{u}(1) = i\,\mathbf{R}$, the Lie algebra of $U(1) \subset \mathrm{C}(V)$, so we can take the orthonormal basis of $\mathfrak{g}$ to be equal to an orthonormal basis of $\mathfrak{spin}(V)$, together with an element of $\mathfrak{u}(1)$. Since the latter is a scalar, its square is a scalar too.

We choose the inner product $\beta$ in $\mathfrak{spin}(V)$ to be a suitable negative multiple of

$$(a,\, b) \mapsto \text{trace}\,(\tau'(a) \circ \tau'(b))\,,$$

where $\tau$ is the two-fold covering from Spin$(V)$ onto SO$(V)$ defined in (4.2), so $\tau'$ is an isomorphism of Lie algebras. This choice ensures that $\beta$ is conjugacy invariant.

If $(\phi_i)_{i=1}^{2n}$ is an orthonormal basis in $V$, then the

$$\phi_i \cdot \phi_j = \tfrac{1}{2}\,(\phi_i \cdot \phi_j - \phi_j \cdot \phi_i)\,, \quad i < j,$$

form a basis in $\mathfrak{spin}(V)$. We read from (4.1) that $\phi_i \cdot \phi_j$ and $\phi_k \cdot \phi_l$ are orthogonal if $i < j$, $k < l$ and $\{i,\, j\} \neq \{k,\, l\}$. Indeed, if $\{i,\, j\}$ and $\{k,\, l\}$ are disjoint, then the composition of $\tau'\,(\phi_i \cdot \phi_j)$ and $\tau'\,(\phi_k \cdot \phi_l)$ is equal to zero. And if for instance $j = k$, then this composition maps $\zeta$ to $4\,(\phi_l,\, \zeta)\,\phi_i$; this mapping has trace equal to zero. On the other hand, the trace of $\tau'(\phi_i \cdot \phi_j)^2$ is equal to $-8$, so after multiplying $\beta$ by a suitable positive constant, we get that the $\phi_i \cdot \phi_j$, with $i < j$, form an orthonormal basis in $\mathfrak{spin}(V)$.

Now, if $i < j$, we get

$$(\phi_i \cdot \phi_j)^2 = \phi_i \cdot \phi_j \cdot \phi_i \cdot \phi_j = -\phi_i \cdot \phi_i \cdot \phi_j \cdot \phi_j = -(-1) \cdot (-1) = -1,$$

which is a scalar. □

## 11.2 Computations in the Exterior Algebra

Since taking the trace is a linear form, we can bring it under the integral sign. Since also

$$\text{trace}_{\mathbf{C}}\,(A \otimes B) = \text{trace}_{\mathbf{C}}\, A \cdot \text{trace}_{\mathbf{C}}\, B, \tag{11.2}$$

the traces factorize with respect to the action on $E$ and on $L_x$.

Using the formula (3.33) for the supertrace, we see that the only factor in the supertrace coming from End $(E)$ is equal to

$$\mathrm{str}_{\mathbf{C}}\left(\mathrm{e}^{\mathrm{c}(X)}\right)=\left(\tfrac{2}{i}\right)^{n}\,\mathrm{vol}\left(\sigma_{\Lambda}\left(\exp_{\mathrm{C}}X\right)\right). \tag{11.3}$$

Here $\sigma_\Lambda$ is the linear mapping from the Clifford algebra $\mathrm{C}(V)\otimes\mathbf{C}$ to $\Lambda V$, the space of all complex-valued exterior forms on $W$, which was defined in (3.15). The multiplication dot refers to the multiplication in the Clifford algebra, which is also used in the power series

$$\exp_{\mathrm{C}}X=\sum_{k=0}^{\infty}\tfrac{1}{k!}X^{k}$$

in $\mathrm{C}(V)\otimes\mathbf{C}$.

The following lemma is a slight variation of [9, Prop. 3.13].

**Lemma 11.2** *For sufficiently small $T\in\mathrm{End}(V)$, we define*

$$c(T)=\left(\det\cosh\tfrac{1}{2}T\right)^{1/2}\in\boldsymbol{R}, \tag{11.4}$$

*and*

$$H(T)=\left(\frac{\tanh\frac{1}{2}T}{\frac{1}{2}T}\right)^{1/2}\in\mathrm{End}(V). \tag{11.5}$$

*Then, if $X\in\mathfrak{spin}(V)$ is sufficiently small:*

$$\sigma_{\Lambda}\left(\exp_{\mathrm{C}}X\right)=c(\tau'(X))\,\Lambda H(\tau'(X))\left(\exp_{\Lambda}\sigma_{\Lambda}(X)\right). \tag{11.6}$$

*Here $\tau'(X)$ is viewed as a linear mapping from $V$ to itself. For any $A\in\mathrm{End}(V)$, $\Lambda A$ denotes the induced action (of pull-back by means of $A'$) on the exterior algebra $\Lambda V$ of $V$. Finally $\exp_\Lambda$ denotes the exponential power series in the algebra $\Lambda V$.*

**Proof** Using the decomposition of $V$ into planes of infinitesimal rotations for $\tau'(X)$, we get the existence of an orthonormal basis $\phi_j$ in $V$ and $\theta_j\in\mathbf{R}$ such that

$$X=\sum_{j=1}^{n}\theta_j\,\phi_{2j-1}\cdot\phi_{2j}. \tag{11.7}$$

Since $\phi_{2j-1} \cdot \phi_{2j}$ commutes with $\phi_{2k-1} \cdot \phi_{2k}$ in $\mathrm{C}(V)$ if $j \neq k$, we get, also using (4.3), that

$$\exp_{\mathrm{C}} X = \prod_{j=1}^{n} \exp_{\mathrm{C}} \theta_j \, \phi_{2j-1} \cdot \phi_{2j} = \prod_{j=1}^{n} (\cos \theta_j + \sin \theta_j \, \phi_{2j-1} \cdot \phi_{2j}) \, .$$

Since $\mathrm{i}(Q\phi_j)(\phi_k) = 0$ if $j \neq k$, this in turn leads to

$$\sigma_\Lambda \left(\exp_{\mathrm{C}} X\right) = \prod_{j=1}^{n} (\cos \theta_j + \sin \theta_j \, \phi_{2j-1} \wedge \phi_{2j}) \, ,$$

see (3.15) for the definition of $\sigma_\Lambda$.

On the other hand,

$$\sigma_\Lambda(X) = \sum_{j=1}^{n} \theta_j \, \phi_{2j-1} \wedge \phi_{2j}.$$

Since the elements $\phi_{2j-1} \wedge \phi_{2j}$ commute in the algebra $\Lambda V$, and have square equal to zero, we get

$$\exp_\Lambda \sigma_\Lambda(X) = \prod_{j=1}^{n} (1 + \theta_j \, \phi_{2j-1} \wedge \phi_{2j}) \, .$$

If $i \neq j$, we will denote by $J_{i,j}$ the linear transformation in $V$ such that

$$J_{i,j}(\phi_i) = \phi_j, \;\; J_{i,j}(\phi_j) = -\phi_i, \;\; J_{i,j}(\phi_k) = 0 \text{ if } k \neq i, \; k \neq j. \qquad (11.8)$$

Then

$$\tau'(X) = \sum_{j=1}^{n} 2 \, \theta_j \, J_{2j-1,\, 2j},$$

so $\mathrm{e}^{\frac{1}{2}\tau'(X)}$ is equal to the rotation in the $\phi_{2j-1}$, $\phi_{2j}$-plane over the angle $\theta_j$. It follows that $\cosh \frac{1}{2}\tau'(X)$ is equal to multiplication by $\cos \theta_j$ in the $\phi_{2j-1}$, $\phi_{2j}$-plane, and hence

$$\left(\det \cosh \tfrac{1}{2}\tau'(X)\right)^{1/2} = \prod_{j=1}^{n} \cos \theta_j .$$

Furthermore,

$$\frac{\tanh \frac{1}{2}\tau'(X)}{\frac{1}{2}\tau'(X)} = \prod_{j=1}^{n} \frac{\sin \theta_j}{\theta_j \cos \theta_j} J_{2j-1,\,2j},$$

which acts on $\phi_{2j-1} \wedge \phi_{2j}$ as multiplication by

$$\left( \frac{\sin \theta_j}{\theta_j \cos \theta_j} \right)^2 .$$

This proves (11.6). □

If $\phi_j$ is an orthonormal basis in $V$, then we write, for each strictly increasing sequence $J = (j_i)_{i=1}^{p}$,

$$\phi_J = \phi_{j_1} \wedge \phi_{j_2} \wedge \ldots \wedge \phi_{j_p} \in \Lambda^p V. \tag{11.9}$$

Together, the $\phi_J$ form a basis of the vector space $\Lambda V$. A natural inner product in $\Lambda V$ is the one for which the $\phi_J$ form an orthonormal basis. Note that

$$\omega = \sum_J (\omega,\, \phi_J)\, \phi_J \tag{11.10}$$

for any $\omega \in \Lambda V$. Applying this to $\omega = \exp_\Lambda \sigma_\Lambda(X)$, using (11.6), and applying (11.10) again to the resulting elements $\Lambda H(\tau'(X))(\phi_J)$, we get

$$\sigma_\Lambda (\exp_{\mathbf{C}} X) = c(\tau'(X)) \sum_{J,K} p_J\, q_{JK}\, \phi_K, \tag{11.11}$$

in which

$$p_J = (\exp_\Lambda \sigma_\Lambda(X),\, \phi_J)$$

and

$$q_{JK} = (\Lambda H(\tau'(X))\phi_J,\, \phi_K)\, .$$

Note that $\Lambda H(\tau'(X))$ preserves the grading of $\Lambda V$, so we only get contributions when $\phi_J$ and $\phi_K$ have the same degree.

The idea is now to collect all the scalar factors, depending on $X$, bring these in front of $\exp_\Lambda \sigma_\Lambda(X)$, and then perform the integration over $X$. After summing over $J$ and $K$ and taking the volume part, we then end up with the supertrace of (9.34). The integration in (9.34) is over $\mathfrak{g} = \mathfrak{u}(1) \oplus \mathfrak{h}$, in which $\mathfrak{g} = \mathfrak{spin}^c(V)$ and $\mathfrak{h} = \mathfrak{spin}(V)$. If we first perform the integration over $\mathfrak{u}(1)$, taking along a factor $(4\pi t)^{-1/2}$, then we are left with an integral over $\mathfrak{h}$ with the same properties. We now study the integral over $\mathfrak{h}$.

In the sequel, we will make systematic use of the commutative algebra

$$\Lambda^+ V := \sum_{j=0}^{n} \Lambda^{2j} V, \tag{11.12}$$

the even part of the exterior algebra $\Lambda V$ of $V$.

**Lemma 11.3** *Let $f$ be a smooth, compactly supported function on $\mathfrak{h} = \mathfrak{spin}(V)$, in which we take the orthonormal basis $e_r = e_r^{\mathfrak{h}}$, consisting of the $\phi_i \cdot \phi_j$, with $i < j$. We write*

$$\psi_r = \sigma_\Lambda(e_r) = \phi_i \wedge \phi_j$$

*for the corresponding basis of $\Lambda^2 V$. Then we have an asymptotic expansion, for $t \downarrow 0$, of the form*

$$(4\pi t)^{-\frac{1}{2}\dim \mathfrak{h}} \int_{\mathfrak{h}} \mathrm{e}^{-\|X\|^2/4t} f(X) \exp_\Lambda \sigma_\Lambda(X)\, \mathrm{d}X \sim \sum_{k=0}^{\infty} t^k \omega_k, \tag{11.13}$$

*in which $\omega_k \in \sum_{j \leq k} \Lambda^{2j} V$. Moreover, if $0 \leq k \leq n$, then the part of $\omega_k$ of degree $2k$ is given by*

$$\omega_k = \tfrac{1}{k!} \left( 2 \sum_r \psi_r\, \partial_r \right)^k f(0) \mod \sum_{j<k} \Lambda^{2j} V. \tag{11.14}$$

*Here $\partial_r$ denotes the partial derivative with respect to the $r$-th coordinate $X_r$ with respect to the orthonormal basis of the $e_r$'s.*

**Proof** Since

$$(4\pi t)^{-\frac{1}{2}\dim \mathfrak{h}}\, \mathrm{e}^{-\|X\|^2/4t}$$

is the heat kernel for the constant coefficient Laplace operator in $\mathfrak{h}$, we get for (11.13) the asymptotic expansion

$$\sum_{k=0}^{\infty} \frac{t^k}{k!} \left( \sum_r \partial_r{}^2 \right)^k (f(X) \exp_\Lambda \sigma_\Lambda(X)) \,|_{X=0}.$$

Now

$$\partial_r{}^2 \, (f(X) \exp_\Lambda \sigma_\Lambda(X)) = \partial_r{}^2 f(X) \exp_\Lambda \sigma_\Lambda(X)$$
$$+2\psi_r \, \partial_r f(X) \exp_\Lambda \sigma_\Lambda(X) + \psi_r{}^2 \, f(X) \exp_\Lambda \sigma_\Lambda(X).$$

Since

$$\psi_r{}^2 = \phi_i \wedge \phi_j \wedge \phi_i \wedge \phi_j = -\phi_i \wedge \phi_i \wedge \phi_j \wedge \phi_j = 0,$$

this leads to the asymptotic expansion

$$\sum_{k=0}^{\infty} \frac{t^k}{k!} \left( \sum_r \partial_r{}^2 + 2 \sum_r \psi_r \, \partial_r \right)^k f(0).$$

If, for $k \leq n$, one wants to get a form of degree $2k$ in the coefficient of $t^k$, then in each factor

$$\sum_r \partial_r{}^2 + 2 \sum_r \psi_r \, \partial_r,$$

one has to take the summand $2 \sum_r \psi_r \, \partial_r$ because the summand $\sum_r \partial_r{}^2$ does not increase the form degree. □

The formula (11.14) means that in the homogeneous part of degree $k$ of the Taylor expansion of $f$ at $X = 0$, with respect to the basis $\phi_i \cdot \phi_j$, $i < j$, we substitute each coordinate $X_r = (X, \, \phi_i \cdot \phi_j)$ by $2\,\psi_r = 2\,\phi_i \wedge \phi_j$. Or, if we take the sum over $k$ of the terms, which have highest form degree:

$$\sum_{k \geq 0} \frac{t^k}{k!} \left( 2 \sum_r \psi_r \, \partial_r \right)^k f(0),$$

then we get an expansion which is equal to the Taylor series of $f$ at the origin *in which each coordinate $X_r$ has been replaced by $2\,t\,\psi_r$*. Note that the discarding of terms of form degree smaller than $2k$ in the coefficient of $t^k$ gives that the coefficient of $t^k$ belongs to $\Lambda^{2k}V$. In this situation, the substitution of $X_r$ by $2\,t\,\psi_r$ is the same as substitution of the $X_r$ by $2\,\psi_r$ and multiplication of the $\Lambda^{2k}V$-term by $t^k$.

Read in another way, Lemma 11.3 says that, if $k \leq n$, then *the contribution to the asymptotic expansion of the part of* (11.13), *which has form degree equal to* $2k$, *only has terms of order* $t^l$ *with* $l \geq k$. (And the coefficient of $t^k$ is given by (11.14)).

In particular, this already explains why the supertrace of (9.34) converges as $t \downarrow 0$, that is, it has no terms of order $t^k$ with $k < 0$ in its asymptotic expansion. Indeed, according to (11.3) we have to take the volume part, that is the part of form degree equal to $2n$. This means that in (11.11) we have to take $\phi_K$, and hence $\phi_J$ of form degree equal to $2n$. But then the integration over $X \in \mathfrak{h}$ yields a factor $t^n$; the convergence of (9.34) for $t \downarrow 0$ now follows from the observation that

$$t^{-\frac{1}{2}\dim P} = t^{-\frac{1}{2}\dim \mathfrak{g}}\, t^{-n}.$$

Note that $t^{-1/2}$ times the integral over $\mathfrak{u}(1)$ also converges, so it does no harm that in Lemma 11.3 we integrated over $\mathfrak{h}$ instead of over $\mathfrak{g}$.

The following lemmas show that, although the factors $f(X)$ which appear in the computation are quite complicated, the substitution (11.14) leads to considerable simplifications.

**Lemma 11.4** *If* $k \geq 2$, *then each matrix coefficient of* $\tau'(X)^k$ *becomes equal to zero, if in its polynomial expansion each coordinate* $(X,\, \phi_i \cdot \phi_j)$, $i < j$, *is replaced by* $2\,\phi_i \wedge \phi_j$.

**Proof** It suffices to prove the lemma for $k = 2$. Using the notation of (11.8), we have

$$\tau'(X) = \sum_{i<j} 2\,(X,\, \phi_i \cdot \phi_j)\, J_{i,j}.$$

Since $J_{i,j} \circ J_{k,l} = 0$ if $\{i, j\}$ and $\{k, l\}$ are disjoint, we get that

$$\tau'(X)^2 = \sum_{i<j,\, k<l} 4\,(X,\, \phi_i \cdot \phi_j)\,(X,\, \phi_k \cdot \phi_l)\, J_{i,j} \circ J_{k,l},$$

where the sum is only over those combinations for which $\{i, j\}$ and $\{k, l\}$ have an element in common. But then $(\phi_i \wedge \phi_j) \wedge (\phi_k \wedge \phi_l) = 0$. □

The quantities $c(\tau'(X))$ and $\Lambda H(\tau'(X))$ only depend on the $\mathfrak{spin}(V)$-part of $X$ because $\mathfrak{u}(1) = \ker \tau'$. Furthermore, because (11.4) and (11.5) are even power series, we get that $c(\tau'(X)) = 1$ and

$$(\Lambda H(\tau'(X))\, \phi_J,\, \phi_K) = \delta_{JK},$$

if we replace each coordinate $(X,\, \phi_i \cdot \phi_j)$ by $2\,\phi_i \wedge \phi_j$. In other words, in the determination of the leading term of the supertrace of (9.34), we are allowed to replace $\mathrm{e}^{\rho'(X)}$ by $\exp_\Lambda \sigma_\Lambda(X)$.

It remains to determine the value of the factor

$$j_{\mathfrak{g}}(\mathrm{ad}\, X)^{1/2}\, j_M(X \cdot \Omega_p)^{-1/2}\, \mathrm{trace}_{\mathbf{C}} \left(\mathrm{e}^{\frac{1}{2} X \cdot R^1}\right), \tag{11.15}$$

coming from (9.36), where we may restrict $X$ to $\mathfrak{spin}(V)$, and we replace in the Taylor expansion of (11.15) at $X = 0$ each coordinate $(X,\, \phi_i \cdot \phi_j)$, $i < j$, by $2\,\phi_i \wedge \phi_j$.

**Lemma 11.5** $j_{\mathfrak{g}}(\mathrm{ad}X)$ *only depends on the* $\mathfrak{spin}(V)$*-part of* $X$. *Moreover, if we replace each coordinate* $(X,\, \phi_i \cdot \phi_j)$ *by* $2\,\phi_i \wedge \phi_j$, *then* $j_{\mathfrak{g}}(\mathrm{ad}X) = 1$.

**Proof** The first statement is because $\mathrm{ad}\, X = 0$ if $X \in \mathfrak{u}(1)$.

Since the determinant of a matrix $A$ is equal to a polynomial in the traces of the powers of $A$, it is sufficient to prove that $\mathrm{trace}(\mathrm{ad}\, X)^k = 0$, if $k \geq 1$, and each coordinate $(X,\, \phi_i \cdot \phi_j)$ is replaced by $\phi_i \wedge \phi_j$.

If $i \neq j$, then

$$\mathrm{ad}\, X\, (\phi_i \cdot \phi_j) = \tau'(X)(\phi_i) \cdot \phi_j + \phi_i \cdot \tau'(X)(\phi_j),$$

the $\sigma_\Lambda$-image of which is equal to

$$\tau'(X)(\phi_i) \wedge \phi_j + \phi_i \wedge \tau'(X)(\phi_j).$$

In other words, the linear isomorphism $\sigma_\Lambda$, from $\mathfrak{spin}(V)$ onto $\Lambda^2 V$, conjugates $\operatorname{ad} X$ to the restriction to $\Lambda^2 V$ of $\operatorname{der} \tau'(X)$, the derivation of the algebra $\Lambda V$ induced by $\tau'(X) \in \operatorname{End}(V)$.

Since $\sigma_\Lambda$ conjugates $\operatorname{ad} X$ with $\operatorname{der} \tau'(X)$, it conjugates $(\operatorname{ad} X)^k$ with $(\operatorname{der} \tau'(X))^k$. Furthermore,

$$(\operatorname{der} \tau'(X))^k \, (\phi_i \wedge \phi_j)$$

is a linear combination of the $\tau'(X)^l(\phi_i) \wedge \tau'(X)^m(\phi_j)$ with $l + m = k$. So we see from Lemma 11.4 that all the matrix coefficients of $(\operatorname{ad} X)^k$ vanish if $k \geq 3$ and each coordinate $(X,\, \phi_i \cdot \phi_j)$ is replaced by $\phi_i \wedge \phi_j$.

With the notation of (11.8), we have

$$\tau'(X) = \sum_{i<j} 2\,(X,\, \phi_i \cdot \phi_j)\, J_{i,j}.$$

So, if $k < l$, then $\tau'(X)(\phi_k) \wedge \phi_l$ is a linear combination of $\phi_m \wedge \phi_l$ with $m \neq k$, and therefore the trace of the restriction to $\Lambda^2 V$ of $\operatorname{der} \tau'(X)$ is equal to zero.

Finally,

$$\begin{aligned}&(\operatorname{der} \tau'(X))^2 \, (\phi_k \wedge \phi_l)\\ &= \tau'(X)^2(\phi_k) \wedge \phi_l + 2\, \tau'(X)(\phi_k) \wedge \tau'(X)(\phi_l) + \phi_k \wedge \tau'(X)^2(\phi_l).\end{aligned}$$

And $\tau'(X)(\phi_k) \wedge \tau'(X)(\phi_l)$ is equal to $4\,(X,\, \phi_k \cdot \phi_l)^2$ times $\phi_k \wedge \phi_l$, plus a linear combination of $\phi_i \wedge \phi_j$, with $i < j$ and $\{i,\, j\} \neq \{k,\, l\}$. The proof is complete in view of the fact that $(X,\, \phi_k \cdot \phi_l)^2 = 0$ if $(X,\, \phi_k \cdot \phi_l)$ is replaced by $\phi_k \wedge \phi_l$. $\square$

**Lemma 11.6** *Fix an orthonormal frame $f_j$ in the tangent bundle of $M$, which we also use to identify $\mathrm{T}_x^* M$ with $V = (\mathbf{R}^n)^* \simeq \mathbf{R}^n$.*

*If we replace the $\mathfrak{u}(1)$-part of $X$ by zero and each coordinate $(X,\, \phi_i \cdot \phi_j)$ by $2\, \phi_i \wedge \phi_j$, then the matrix*

$$A_{kl} = \Big( (X \cdot \Omega_p)\, f_{l,\mathrm{hor}},\, f_{k,\mathrm{hor}} \Big)$$

*of $X \cdot \Omega_p$, cf. (9.4), turns into the $\Lambda^2 \mathrm{T}_x^* M$-valued matrix*

$$(v,\, w) \mapsto - \Big( R^\beta(v,\, w)\, f_l,\, f_k \Big)\,.$$

**Proof** We may assume that $X \in \mathfrak{spin}(V)$. Since the connection in $\mathrm{Spin}^{\mathrm{c}}\,\mathrm{F}\,M$ is related to the Levi-Civita connection in $\mathrm{SOF}\,M$ via the mapping $\alpha$ in (5.3), and we have the relation (6.9) between curvature forms and curvature operators, we get

$$A_{kl} = - \Big( \tau'(X),\, R^\beta(f_l,\, f_k) \Big)\,,$$

where the inner product in $\mathfrak{so}(V)$ is the push-forward of the inner product in $\mathfrak{spin}(V)$ under $\tau'$. In terms of (6.5) and (11.8), we have

$$R^\beta(f_l,\, f_k) = \sum_{i<j} R^\beta_{jilk}\, J_{i,j}\,,$$

whereas on the other hand

$$\tau'(X) = \sum_{i<j} (X,\, \phi_i \cdot \phi_j)\; 2\, J_{i,j}\,,$$

so the $2\, J_{i,j}$ with $i < j$ form an orthonormal basis in $\mathfrak{so}(V)$. Therefore,

$$A_{kl} = -\tfrac{1}{2} \sum_{i<j} (X,\, \phi_i \cdot \phi_j)\; R^\beta_{jilk}.$$

Because of the antisymmetry of the curvature and the relation (6.18), we have

$$R^\beta_{jilk} = R^\beta_{ijkl} = R^\beta_{klij}.$$

This completes the proof, because

$$(\phi_i \wedge \phi_j)(f_i,\, f_j) = 1. \tag{11.16}$$

if the $\phi_j$ form a dual basis to the $f_j$. □

**Lemma 11.7** *Fix an orthonormal frame $f_j$ in $TM$, with dual frame $\phi_j$, which is used to identify $\mathrm{T}^*_xM$ with $V = \mathbf{R}^n$. Also fix a unitary frame $l_r$ in the vector bundle $L$.*

*If we replace the $\mathrm{u}(1)$-part of $X$ by zero and each coordinate $(X,\, \phi_i \cdot \phi_j)$ by $2\,\phi_i \wedge \phi_j$, then the matrix*

$$A_{rs} = \left(\left(\tfrac{1}{2}\, X \cdot R^1\right)(l_s),\, l_r\right)$$

*of $\frac{1}{2}\, X \cdot R^1$, cf. (9.32), turns into the $\Lambda^2\mathrm{T}^*_xM$-valued matrix*

$$(u,\, v) \mapsto \tfrac{1}{2}\, R^{K^*}(u,\, v) + \left(R^L(u,\, v)\, l_s,\, l_r\right).$$

*Here $R^{K^*}$ is viewed as an $i\,\mathbf{R}$-valued two-form on $M$.*

**Proof** If we compare (9.18) with (6.10) and (6.11), then we see that (9.32) takes the form

$$X \cdot R^1 = \sum_{i<j} (X,\, \phi_i \cdot \phi_j)\, \left[\tfrac{1}{2}\, R^{K^*}(f_i,\, f_j) + R^L(f_i,\, f_j)\right].$$

The result follows in view of (11.16). □

## 11.3 The Short Time Limit of the Supertrace

We have now collected all the ingredients for the computation of the short time limit of the supertrace of the heat kernel along the diagonal.

**Theorem 11.1** *Let $(M,\, J)$ be an almost complex manifold. Let $\beta$ be a Riemannian structure in $M$, with curvature $R^{\beta}$, which at each point $x \in M$ is viewed as an element of $\Lambda^2 \mathrm{T}_x^* M \otimes \mathrm{End}(\mathrm{T}_x M)$. Let the dual $K^*$ of the canonical line bundle $K = \mathrm{T}^* M^{(n,\,0)}$ be provided with a Hermitian connection, the curvature $R^{K^*}$ of which is viewed as an $i\,\mathbf{R}$-valued two-form on $M$. Finally, let $L$ be a complex vector bundle over $M$, provided with a Hermitian connection, of which the curvature $R^L$ at $x \in M$ is viewed as an element of $\Lambda^2 \mathrm{T}_x^* M \otimes \mathrm{End}(L_x)$.*

*Let $D$ be the spin-c Dirac operator, defined in (5.14). Let $K(t,\, x,\, y)$ be the asymptotic kernel of the operator $\mathrm{e}^{-t\, D^2}$, as discussed after Theorem 8.1. Denote by $K^{\pm}(t,\, x,\, y)$ the restriction of $K(t,\, x,\, y)$ to $E^{\pm} \otimes L$. Then, for each $x \in M$, the volume form*

$$\left(\mathrm{trace}_{\mathbf{C}} K^+(t,\, x,\, x) - \mathrm{trace}_{\mathbf{C}} K^-(t,\, x,\, x)\right)\, \mathrm{d}x$$

*converges for $t \downarrow 0$ to the volume part of*

$$(2\pi i)^{-n} \cdot \left(\det \frac{1 - \mathrm{e}^{-R^{\beta}}}{R^{\beta}}\right)^{-1/2} \cdot \mathrm{e}^{\frac{1}{2}\, R^{K^*}} \cdot \mathrm{trace}_{\mathbf{C}} \left(\mathrm{e}^{R^L}\right). \tag{11.17}$$

*Here the determinant and the trace are taken of matrices, the coefficients of which are elements of $\Lambda^2 \mathrm{T}_x^* M$. These belong to the commutative algebra $\Lambda_x^+ = \Lambda^{\mathrm{even}} \mathrm{T}_x^* M$, so (11.17) is an element of $\Lambda_x^+$.*

*If $M$ is compact, then the integral over $M$ of the volume part of (11.17) is equal to the index (2.39) of the spin-c Dirac operator, which in the complex analytic case in turn is equal to the Riemann-Roch number (2.20).*

**Proof** If we take the supertrace of (9.34) and apply (11.2), Lemma 11.1, and the fact that $\dim P = 2n + \dim \mathfrak{g}$, then we get $(4\pi t)^{-n}\, \mathrm{e}^{-t\, \mathrm{Cas}}$ times

$$(4\pi t)^{-\frac{1}{2}\, \dim \mathfrak{g}} \int_{\mathfrak{g}} \mathrm{e}^{-\|X\|^2/4t}\, \mathrm{str}_{\mathbf{C}} \left(\mathrm{e}^{\mathrm{c}(X)}\right) \sum_{k=0}^{\infty} t^k\, \mathrm{trace}_{\mathbf{C}}\, B_k(p,\, X)\, \mathrm{d}X, \tag{11.18}$$

in which $B_k(p, X) \in \mathrm{End}(L_x)$, and $B_0$ is given by (9.36).

The factor $(\frac{2}{i})^n$ in (11.3) changes the factor $(4\pi t)^{-n}$ into $(2\pi i)^{-n}$ times $t^{-n}$. Using Lemma 11.2, Lemma 11.3 and Lemma 11.4, we get that the leading term of (11.18), which corresponds to the volume part in (11.13), is equal to $t^n$ times the volume part of (11.15). In (11.15) we take $X \in \mathfrak{spin}(V)$, and replace, in the Taylor expansion of (11.15) at $X = 0$, each coordinate $(X, \phi_i \cdot \phi_j)$, $i < j$, by $2\,\phi_i \wedge \phi_j$. Note that the factor $\mathrm{e}^{-t\,\mathrm{Cas}}$ drops out of the computation of the leading term. The formula (11.17) now follows from Lemma 11.5, Lemma 11.6, and Lemma 11.7. □

The volume part of (11.17) is a polynomial expression in the matrix coefficients of $R^\beta$, $R^{K^*}$ and $R^L$, which are two-forms on $M$, so the polynomial expression is homogeneous of degree $n$. As a consequence, the volume part of (11.17) remains the same if in (11.17) we delete the factor $(2\pi i)^{-n}$, and replace $R^\beta$, $R^{K^*}$ and $R^L$ by $R^\beta/2\pi i$, $R^{K^*}/2\pi i$ and $R^L/2\pi i$, respectively. This is natural in the translation of the volume part of (11.17) in terms of characteristic classes. See Chapter 13, especially Proposition 13.1.

# Chapter 12

# The Local Lefschetz Fixed Point Formula

In this chapter we will discuss the modifications of the previous chapter, which are needed for the computation of the leading term of the supertrace of (10.18) in the presence of an automorphism $\gamma$. See formula (12.12) in Theorem 12.1; in the Kähler case the answer can be brought in the simpler form of Lemma 12.3. The translation of the formulas in terms of characteristic classes will be given in Proposition 13.2.

## 12.1 The Element $g_0$ of the Structure Group

It will be convenient to write

$$\mathrm{e}^{\mathrm{c}(X)} \circ \mathrm{c}(g_0) = \mathrm{c}(g_0) \circ \mathrm{e}^{\mathrm{c}\left(\mathrm{Ad}\, g_0{}^{-1}(X)\right)}, \tag{12.1}$$

and replace the integration variable $X$ by $\mathrm{Ad}\, g_0(X)$. Writing $n_F = n - l_F$, we have

$$\dim F = 2\, n_F, \tag{12.2}$$

and

$$\dim P = 2\, l_F + 2\, n_F + \dim \mathfrak{g}.$$

So we have to determine the leading term in the asymptotic expansion for $t \downarrow 0$, of the product of $\mathrm{e}^{-t\,\mathrm{Cas}}\,(4\pi t)^{-n_F}$ with the supertrace of

$$(4\pi t)^{-\frac{1}{2}\dim\mathfrak{g}} \int_{\mathfrak{g}} \mathrm{e}^{-\|X\|^2/4t} \sum_{k=0}^{\infty} t^k\, \tilde{C}_k(p,\, X) \circ \mathrm{c}(g_0) \circ \mathrm{e}^{\mathrm{c}(X)}\, \mathrm{d}X. \qquad (12.3)$$

Here

$$\tilde{C}_k(p,\, X) = C_k\left(p,\, \mathrm{Ad}\, g_0(X)\right)$$

and the $C_k$ are as in (10.19) and (10.20). Before we continue, we check that:

**Lemma 12.1** *The coefficient $C_0$ satisfies the invariance property that*

$$C_0(p,\, \mathrm{Ad} g_0(X)) = C_0({g_0}^{-1}p,\, X).$$

**Proof** We see from

$$[\mathrm{Ad}\, g_0(X),\, Y] = \mathrm{Ad}\, g_0\, \left[X,\, \mathrm{Ad}\, {g_0}^{-1}(Y)\right]$$

that

$$\mathrm{ad}\,(\mathrm{Ad}\, g_0(X)) = \mathrm{Ad}\, g_0 \circ \mathrm{ad}\, X \circ \mathrm{Ad}\, {g_0}^{-1}.$$

Hence

$$J_{\mathfrak{g}}\,(\mathrm{ad}\,(\mathrm{Ad}\, g_0(X))) = \mathrm{Ad}\, g_0 \circ J_{\mathfrak{g}}\,(\mathrm{ad}\, X) \circ \mathrm{Ad}\, {g_0}^{-1},$$

which has the same determinant as $J_{\mathfrak{g}}\,(\mathrm{ad}\, X)$.

Secondly, the definition (9.4) of $X \cdot \Omega_p$ and the $G$-equivariance of the curvature form $\Omega$ imply that

$$\mathrm{Ad}\, g_0(X) \cdot \Omega_p = \mathrm{T}_{{g_0}^{-1}p}\, g_0 \circ (X \cdot \Omega_{{g_0}^{-1}p}) \circ \left(\mathrm{T}_{{g_0}^{-1}p}\, g_0\right)^{-1}.$$

We use that $\gamma_p$ commutes with $g_0$ and that the determinant is conjugacy-invariant.

Finally we have that

$$\left(\operatorname{Ad} g_0(X) \cdot R^1\right)(p) = \left(X \cdot R^1\right)\left({g_0}^{-1} p\right),$$

because of (9.32) and the invariance property (9.19). □

As in (11.3), we get that the only factor in the supertrace coming from End (E($n$)) is equal to

$$\operatorname{str}_{\mathbf{C}}\left(\mathrm{c}(g_0) \circ \mathrm{e}^{\mathrm{c}(X)}\right) = \left(\tfrac{2}{i}\right)^n \operatorname{vol}\left(\mathrm{c}_\Lambda(g_0)\, \sigma_\Lambda\left(\exp_{\mathrm{C}} X\right)\right). \tag{12.4}$$

It was for the purpose of getting the element (11.6) in the formula that we moved $\mathrm{c}(g_0)$ over to the left. We now investigate the action $g_{0,P}$ and $\mathrm{c}_\Lambda(g_0)$ of $g_0$ on $P$ and $\Lambda V$, respectively.

Recall that $g_0$ was defined as the element of $G$, the group $\mathrm{Spin}^{\mathrm{c}}(2n)$, such that $g_0\, \gamma_P(p_0) = p_0$, cf. the text preceding (10.9). Note that we still have the freedom to choose $p_0$ in the fiber $P_x$ of $P$ over the fixed point $x$. In view of the definition (5.1) of the spin-c frame bundle $P$, it natural to choose a unitary frame $e$ in $\mathrm{T}_x\, M$ and then take $p_0$ equal to the $\mathrm{U}(n)$-orbit of $(e,\, 1)$ in the Cartesian product of UF $M$ and $\mathrm{Spin}^{\mathrm{c}}(2n)$. In other words, we take $p_0$ in UF $M$, viewed as a subbundle of $P$.

The transformation $\gamma_P$ leaves UF $M$ invariant, its restriction to UF $M$ is equal to the action of $\mathrm{T}\,\gamma$ on UF $M$. Since $g_0$ acts on UF $M \times \mathrm{Spin}^{\mathrm{c}}(2n)$ as right multiplication of the second factor by ${g_0}^{-1}$, the equation $g_0\, \gamma_P(p_0) = p_0$ now means that there exists an $A \in \mathrm{U}(n)$ such that

$$\left(\mathrm{T}_x\, \gamma \circ e,\, {g_0}^{-1}\right) = \left(e \circ A^{-1},\, \mathrm{c}^{-1} \circ \Lambda(A)\right).$$

That is,

$$\mathrm{T}_x\, \gamma = e \circ A^{-1} \circ e^{-1}$$

and

$${g_0}^{-1} = \mathrm{c}^{-1} \circ \Lambda(A) \in U.$$

In view of (4.13), this also yields that

$$\tau(g_0) = A^{-1} = e^{-1} \circ \mathrm{T}_x \gamma \circ e.$$

We now make an even more special choice for $e$, namely such that its elements form a basis of eigenvectors of $\mathrm{T}_x \gamma$, viewed as a complex linear transformation, and such that the first $n - l$ elements correspond to the eigenvalue 1. We denote the $j$-th element of the unitary frame $e$ by $e_j \in \mathrm{T}_x M$.

Recall that the connected component $F$ of $M^\gamma$ through $x$ is a smooth manifold of codimension $2l$. The choice of $e$ now means that $e_j$ for $1 \leq j \leq n - l$ form a unitary basis of $\mathrm{T}_x F$, whereas the remaining ones form a unitary basis of the orthogonal complement $N_x$ of $\mathrm{T}_x F$ in $\mathrm{T}_x M$. For each $j$ with $n - l + 1 \leq j \leq n$, there is a $\lambda_j$ on the unit circle in $\mathbf{C}$, such that $\lambda_j \neq 1$ and $\mathrm{T}_x \gamma(e_j) = \lambda_j e_j$.

In other words, $e$ identifies the subspace $W_0$ of $\mathbf{C}^n$, spanned by the first $n - l$ standard base vectors, with $\mathrm{T}_x F$, and the subspace $W_1$ spanned by the remaining ones with $N_x$. With a somewhat dangerous notation, we will write $F$ instead of $W_0$ and $N$ instead of $W_1$. (In view of the normal coordinates which we used in Chapter 10, this notation is only moderately dangerous.) Moreover, $\tau(g_0)$ is diagonal now, with the first $n - l$ diagonal entries equal to 1 and the last $l$ diagonal entries equal to the $\lambda_j$.

Let $\theta_j$ be the unique angle such that

$$\lambda_j = \mathrm{e}^{i\,2\,\theta_j} \quad \text{and} \quad 0 < \theta_j < \pi. \tag{12.5}$$

In view of (4.6) and (4.1), we get that $g_0 = \exp_{\mathbf{C}} X_0$, with

$$X_0 = \sum_{j=n-l+1}^{n} \theta_j \left(\epsilon_j \cdot J\epsilon_j + i\right),$$

if the $\epsilon_j$ form the standard unitary basis of $V$. Therefore, using (4.3), we obtain that

$$g_0 = \prod_{j=n-l+1}^{n} \left(\cos\theta_j + (\sin\theta_j)\, \epsilon_j \cdot J\epsilon_j\right) \mathrm{e}^{i\theta_j}.$$

From this, we see that, if $\omega \in \Lambda^{2i}V$, then

$$c_\Lambda(g_0)\omega = a_\gamma \operatorname{vol}_N \wedge \omega \mod \sum_{h<i+l} \Lambda^{2h}V, \tag{12.6}$$

in which

$$a_\gamma := \prod_{j=n-l+1}^{n} e^{i\theta_j} \sin\theta_j \tag{12.7}$$

and

$$\operatorname{vol}_N := \bigwedge_{j=n-l+1}^{n} \epsilon_j \wedge J\epsilon_j \in \Lambda^{2l}V \tag{12.8}$$

denotes the standard volume in the $2l$-dimensional space $N$.

This leads to the following variation of Lemma 11.3:

**Lemma 12.2** *Let $f$ be a smooth, compactly supported function on $\mathfrak{h} = \mathfrak{spin}(V)$, in which we take the orthonormal basis $e_p^{\mathfrak{h}}$, consisting of the $\phi_i \cdot \phi_j$, with $i < j$. We write $\psi_p = \phi_i \wedge \phi_j$ for the corresponding basis of $\Lambda^2 V$. Then we have an asymptotic expansion, for $t \downarrow 0$, of the form (11.13) in which now $\omega_k \in \sum_{j \leq k+l} \Lambda^{2j}V$. Moreover, if $0 \leq k \leq n-l$, then the part of $\omega_k$ of degree $2k+2l$ is given by*

$$\omega_k = a_\gamma \operatorname{vol}_N \wedge \tfrac{1}{k!} \left(2 \sum_p \psi_p \, \partial_p\right)^k f(0) \text{ modulo } \sum_{j<k} \Lambda^{2j}V. \tag{12.9}$$

*Here $a_\gamma$ is as in (12.7), and $\partial_p$ denotes the partial derivative with respect to the $p$-th coordinate, with respect to the orthonormal basis consisting of the $e_p^{\mathfrak{h}}$.*

## 12.2 The Short Time Limit

Lemma 12.2 implies that the volume part of (11.13) this time has an asymptotic expansion in $t^k$, with $k \geq n - l_F = n_F$, which neatly kills the singular factor $t^{-n_F}$. Therefore, the integral over the diagonal in $N_x \times N_x$, of the supertrace of the integral kernel of (10.4), converges as $t \downarrow 0$.

In order to determine the limit, we use the splitting of $\mathrm{T}_x^* M$ into $\mathrm{T}_x^* F$ and $N_x{}^*$. If $\omega \in \Lambda^r \mathrm{T}_x^* M$, then we write

$$\omega_F := i^*\omega \in \Lambda^r \mathrm{T}_x^* F \tag{12.10}$$

for the pull-back (restricition) of $\omega$ to $\mathrm{T}_x F$. Then $\omega - \omega_F$ is equal to a sum of $\mu \wedge \nu$, with $\mu \in \Lambda^{r-s} \mathrm{T}_x^* F$, $\nu \in \Lambda^s N_x{}^*$, and $s > 0$. In particular, we have for volume forms:

$$\alpha_\gamma \wedge \omega = \alpha_\gamma \wedge \omega_F \quad if\ \omega \in \Lambda^{2n-2l} \mathrm{T}_x^* M. \tag{12.11}$$

On the other hand, the description in Section 10.1 of the fixed point manifold $F$ in normal coordinates shows that the curvature operator $R^\gamma(u, v)$ leaves the subspace $\mathrm{T}_x F$, and therefore also its orthogonal complement $N_x$, invariant.

We have now made all the preparations for:

**Theorem 12.1** *In addition to the assumptions of Theorem 11.1, let $\gamma$ be a transformation in $M$, which preserves the almost complex structure $J$, the Riemannian structure $\beta$, and the connection in $K^*$, and which is covered by an automorphism $\gamma_L$ of $L$, which preserves the Hermitian structure and connection in $L$.*

*Let $K_\gamma^\pm(t, x, y)$ be the asymptotic kernel of the operator $\gamma_{\mathrm{op}} \circ \mathrm{e}^{-t D^2}$, restricted to $E^\pm \otimes L$. Then $K_\gamma^\pm(t, x, x)$ is rapidly decreasing as $t \downarrow 0$, unless $x$ is a fixed point of $\gamma$, so we assume in the sequel that $x$ belongs to the connected component $F$ of $M^\gamma$.*

*Let $\psi$ be a smooth real-valued function, with compact support, contained in a small normal coordinate neighborhood $U$ around $x$ in $M$. Let $N_x$ denote the orthogonal complement of the tangent space at $x$ of $F \cap U$. Then the integral over the diagonal in $N_x \times N_x$ of*

$$\left(\mathrm{trace}_{\mathbf{C}} K_\gamma^+(t, x, x) - \mathrm{trace}_{\mathbf{C}} K_\gamma^-(t, x, x)\right) \mathrm{d}_F x$$

*converges for $t \downarrow 0$ to the $F$-volume part of*

$$(2\pi i)^{-n_F} \cdot \left(\det \frac{1 - \mathrm{e}^{-R^{\beta}_F}}{R^{\beta}_F}\right)^{-1/2} \cdot \mathrm{e}^{\frac{1}{2} R^{K^*_F}} \cdot \mathrm{trace}_{\mathbf{C}} \left(\gamma_{L,x}\, \mathrm{e}^{R^{L_F}}\right)$$
$$\det \left(1 - \gamma_N^{-1}\, \mathrm{e}^{R^{\beta}_N}\right)^{-1/2} \cdot i^{-l} \cdot (\det_{\mathbf{C}} \gamma_N)^{1/2} \cdot \mathrm{e}^{\frac{1}{2} R^{K^*_N}}. \qquad (12.12)$$

*Here we have used the following notations.* $\dim F = 2\, n_F = 2 \dim M - 2\, l$, $l = l_F$. *$R^{\beta}_F$ is the curvature of $(F, \beta)$. $K^*_F$ is the dual of the canonical line bundle of $(F, J)$. $L_F$ is the restriction to $F$ of the vector bundle $L$. $\gamma_N$ is the restriction of $T_x \gamma$ to $N_x$. In the complex determinant, $\gamma_N$ is viewed as a complex-linear transformation in $(N_x, J)$. $R^{\beta}_N$ is the restriction to $N_x$ of $R^{\beta}$, viewed as an element of*

$$\Lambda^2 \mathrm{T}^*_x F \otimes \mathrm{End}(N_x).$$

*Finally, $K^*_N$ is the dual of the complex line bundle of the $(l, 0)$-forms on the normal bundle $N = N(F)$ of $F$ in $M$.*

*If $M$ is compact, then the sum over the connected components $F$ of $M^{\gamma}$, of the integral over $F$ of the $F$-volume part of (12.12), is equal to the virtual character (2.40), which in the complex analytic case is equal to the holomorphic Lefschetz number (2.21).*

**Proof** The arguments are similar to the proof of Theorem 11.1, where however this time all forms coming from the curvatures are restricted to $\mathrm{T}_x F$. We have used that, over $F$, $K^* = K^*_F \otimes K^*_N$, hence

$$R^{K^*} = R^{K^*_F} + R^{K^*_N},$$

which in turn implies that, still over $F$,

$$\mathrm{e}^{\frac{1}{2} R^{K^*}} = \mathrm{e}^{\frac{1}{2} R^{K^*_F}} \cdot \mathrm{e}^{\frac{1}{2} R^{K^*_N}}.$$

We now discuss the factor $\det(1-\gamma_N)^{-1/2}$ in (10.20). In terms of (12.5), the real determinant of $1-\gamma_N$ is equal to

$$\prod_{j=n-l+1}^{n}\left[(1-\cos 2\theta_j)^2+(\sin 2\theta_j)^2\right]$$
$$=\prod_{j=n-l+1}^{n}(2-2\cos 2\theta_j))=\prod_{j=n-l+1}^{n}4\sin^2\theta_j,$$

so

$$\det(1-\gamma_N)^{-1/2}=1/2^l\prod_{j=n-l+1}^{n}\sin\theta_j.$$

Here we have used the choice that $0<\theta_j<\pi$.

Combined with the factors $(\frac{2}{i})^n$ in (12.4) and $a_\gamma$ in (12.6), this leads to a factor $(\frac{2}{i})^{n_F}$, which is used in order to replace the factor $(4\pi)^{-n_F}$ by $(2\pi i)^{-n_F}$, times the factor

$$i^{-l}\prod_{j=n-l+1}^{n}e^{i\theta_j}=i^{-l}\left(\det_{\mathbf{C}}\gamma_N\right)^{1/2}.$$

Here the square root of the complex determinant of $\gamma_N$, viewed as a complex-linear transformation in $(N,\ J)$, is actually *defined* by this formula, in which the $\theta_j$ are chosen as in (12.5) and the $\lambda_j$ are equal to the eigenvalues of $\gamma_N$. Its square is equal to $\det_{\mathbf{C}}\gamma_N$. Moreover, if $\tilde{U}$ denotes the set of unitary transformations in $N$ which have no eigenvalue equal to one, then $\det_{\mathbf{C}}^{1/2}$ is a continuous function on $\tilde{U}$, which converges to 1 as $\gamma_N\to 1$. These properties determine the function $\det_{\mathbf{C}}^{1/2}$ uniquely. □

In the same way as remarked after Theorem 11.1, the $F$-volume part of (11.17) remains the same if in (12.12) we delete the factor $(2\pi i)^{-n_F}$, and replace $R_F^\beta$, $R^{K_F^*}$, $R^{L_F}$, $R_N^\beta$ and $R^{K_N^*}$ by $1/2\pi i$ times these curvatures. Again, this is the natural form in the translation of the volume part of (12.12) in terms of characteristic classes in Chapter 13, cf. Proposition 13.2.

## 12.3 The Kähler Case

If $(M,\, J,\, h)$ is a Kähler manifold, then not only the proof simplifies considerably because we can replace the spin-c group everywhere by the much simpler unitary group, but also the answers get a somewhat more compact form:

**Lemma 12.3** *Let $\tilde{R}^\beta$ be the curvature of a connection $\tilde{\nabla}$ in $\mathrm{T}M$, which leaves both $\beta$ and the almost complex structure $J$ invariant, and let the connection in $K^*$ be defined by $\tilde{\nabla}$. With respect to a unitary frame in $\mathrm{T}M$, $\tilde{R}^\beta$ is viewed as an $n \times n$-matrix of two-forms in $M$. Then*

$$\left(\det \frac{1 - \mathrm{e}^{-\tilde{R}^\beta}}{\tilde{R}^\beta}\right)^{-1/2} \cdot \mathrm{e}^{\frac{1}{2}\, R^{K^*}} = \det_{\mathbf{c}} \frac{\tilde{R}^\beta}{1 - \mathrm{e}^{-\tilde{R}^\beta}}. \tag{12.13}$$

*Furthermore, the quantity on the second line in (12.12) then is equal to*

$$\det_{\mathbf{c}} \left(1 - {\gamma_{_N}}^{-1}\, \mathrm{e}^{-\tilde{R}^\beta_N}\right)^{-1}. \tag{12.14}$$

**Proof** For each $x \in M$ and $v,\, w \in \mathrm{T}_x\, M$, the curvature operator $\tilde{R}^\beta_x(v,\, w)$ is complex linear with respect to $J_x$. And, as observed before in the proof of Proposition 6.1,

$$R^{K^*}(v,\, w) = \mathrm{trace}_{\mathbf{c}}\, \tilde{R}^\beta(v,\, w). \tag{12.15}$$

If $A \in \mathrm{End}\,(\mathrm{T}_x\, M)$ is complex linear with respect to $J$, with correponding complex determinant $\det_{\mathbf{c}} A$, then its real adjoint $A'$ with respect to $\beta_x$ satisfies

$$\det_{\mathbf{c}} A' = \overline{\det_{\mathbf{c}} A}.$$

So the complex conjugate of

$$j_{\mathbf{c}} := \det_{\mathbf{c}} \frac{1 - \mathrm{e}^{-\tilde{R}^\beta}}{\tilde{R}^\beta}$$

is equal to

$$\det_{\mathbf{C}} \frac{e^{\tilde{R}^\beta} - 1}{\tilde{R}^\beta} = \det_{\mathbf{C}} e^{\tilde{R}^\beta} \, \jmath\mathbf{c} = e^{R^{K^*}} \, \jmath\mathbf{c},$$

where in the second identity we have used (12.15). The formula (12.13) now follows because

$$\det \frac{1 - e^{-\tilde{R}^\beta}}{\tilde{R}^\beta} = \jmath\mathbf{c} \cdot \overline{\jmath\mathbf{c}}.$$

For the second statement, we observe that the complex conjugate of the complex determinant of

$$A = 1 - \gamma_N \, e^{-\tilde{R}^\beta_N}$$

is equal to the complex determinant of

$$B = 1 - e^{R^N} \, \gamma_N^{-1} = -\gamma_N^{-1} \, e^{\tilde{R}^\beta_N} \, A.$$

(See also the argument leading to (10.25).) So, because the real determinant of $B$ is equal to the product of the complex determinants of $A$ and $B$:

$$\det B \, \det_{\mathbf{C}} \left( -\gamma_N \, e^{-\tilde{R}^\beta_N} \right) = \left( \det_{\mathbf{C}} A \right)^2.$$

This leads to (12.14), where for the determination of the sign we argue as follows. For each eigenvalue $e^{-2i\theta_j}$ of $\gamma_N$, with $0 < \theta_j < \pi$, the positive square root of the product of $1 - e^{2i\theta_j}$ and its complex conjugate is equal to

$$2 \sin \theta_j = \tfrac{1}{i} \left( e^{i\theta_j} - e^{-i\theta_j} \right).$$

If we multiply this with $i \, e^{-i\theta_j}$, we get $1 - e^{-2i\theta_j}$. □

# Chapter 13

# Characteristic Classes

## 13.1 Weil's Homomorphism

Let $G$ be a Lie group with Lie algebra $\mathfrak{g}$ and let $P$ be a principal $G$-bundle over a manifold $M$, with projection $\pi : P \to M$. As in the paragraph preceding (6.8), let $\theta$ be a connection form in $P$, with curvature form $\Omega$.

We begin with the formulation in Theorem 13.1 of a basic result of André Weil [74], which states that there is a canonical homomorphism, which assigns to each conjugacy invariant polynomial on $\mathfrak{g}$ a de Rham cohomology class of $M$, which, roughly speaking, is obtained by substituting the curvature form in the polynomial. The resulting cohomology classes in $M$ are the characteristic classes in the title of this chapter.

Since the traces and determinant in (11.17) are conjugacy invariant polynomials, this leads to a translation in terms of characteristic classes, which is the usual form for the Hirzebruch-Riemann-Roch theorem. See Proposition 13.1.

In the case of an automorphism $\gamma$, the normal bundle of each connected component of the fixed point set splits into smooth eigenbundles, with constant eigenvalues for the action of the tangent map of $\gamma$. See Lemma 13.1. This is used as a preparation for the translation in Proposition 13.2, of formula

(12.12) and Lemma 12.3, in terms of characteristic classes.

**Theorem 13.1** *Let* $e_i = e_i^{\mathfrak{g}}$ *be a basis in* $\mathfrak{g}$*, write* $X_i$ *for the* $i$*-th coordinate of* $X \in \mathfrak{g}$*, with respect to the basis* $e_j^{\mathfrak{g}}$*. The curvature form is a* $\mathfrak{g}$*-valued two-form in* $P$*, the* $i$*-th coordinate of which is a scalar-valued two-form* $\Omega_i$ *in* $P$*. Let*

$$f(X) = \sum_{\mu} c_\mu \prod_i X_i{}^{\mu_i}$$

*be an* Ad $G$*-invariant complex-valued polynomial on* $\mathfrak{g}$*. Then there is a uniquely defined element*

$$f(\Omega) \in \sum_j \Omega^{2j}(M)$$

*in the commutative algebra* $\Omega^+(M)$ *of sums of differential forms in* $M$ *of even degree, such that*

$$\sum_{\mu} c_\mu \prod_i \Omega_i{}^{\mu_i} = \pi^* \left( f(\Omega) \right).$$

*The form* $f(\Omega)$ *does not depend on the choice of the basis in* $\mathfrak{g}$*. For each* $j$*, the component* $f(\Omega)^{(2j)}$ *of degree* $2j$ *is a closed differential form in* $M$*. If* $\tilde{\theta}$ *is another connection form in* $P$ *with curvature form* $\tilde{\Omega}$*, then there is a differential form* $\beta$ *in* $M$ *of odd degree, such that*

$$f(\tilde{\Omega}) - f(\Omega) = \mathrm{d}\beta. \tag{13.1}$$

*In other words, the de Rham cohomology class*

$$[f(\Omega)] \in \mathrm{H}^+(M) := \sum_j \mathrm{H}^{2j}(M)$$

*of* $f(\Omega)$ *does not depend on the choice of the connection form* $\theta$*.*

The mapping

$$f \mapsto [f(\Omega)] : \mathbf{C}[\mathfrak{g}]^{\mathrm{Ad}\, G} \to \mathrm{H}^+(M) \tag{13.2}$$

is a homomorphism, from the commutative algebra $\mathbf{C}[\mathfrak{g}]^{\mathrm{Ad}\,G}$ of the Ad $G$-invariant polynomials on $\mathfrak{g}$ to the commutative algebra $\mathrm{H}^+(M)$ of sums of de Rham cohomology classes of even degree. This homomorphism is known as the *Weil homomorphism*. The $[f(\Omega)]$, $f \in \mathbf{C}[\mathfrak{g}]^{\mathrm{Ad}\,G}$, are called the *characteristic classes* of the principal $G$-bundle $P$. It is therefore natural to call the differential forms $f(\Omega)$ the *characteristic forms* of the bundle $P$ with connection form $\theta$. (The characteristic forms depend on the choice of the connection form, whereas the characteristic classes don't.)

We will give a proof of Theorem 13.1 in Section 16.4, by showing that it follows from a theorem of H. Cartan, Theorem 16.1, which is formulated in the somewhat more general framework of equivariant differential forms. The proof of Theorem 16.1 also yields an explicit algebraic construction of the form $\beta$ in (13.1).

## 13.2 The Chern Matrix and the Riemann-Roch Formula

Now let $W$ be a complex vector bundle over $M$, with fiber dimension equal to $d$, and suppose that $W$ is provided with a Hermitian structure $h$. Then, for each $x \in M$, a *unitary frame* $e_x$ in $W_x$ is defined as $\mathbf{C}$-basis $\left(e_{x,j}\right)_{j=1}^{d}$ of $W_x$, which is orthonormal with respect to the Hermitian inner product $h_x$ in $W_x$. We will identify the frame $e$ with the complex linear mapping

$$e : (z_1, z_2, \ldots, z_d) \mapsto \sum_{j=1}^{d} z_j\, e_{x,j} : \mathbf{C}^d \to W_x .$$

In this way, the unitary frames in $W_x$ form the set $\mathrm{UF}_x\, W$ of unitary mappings from $\mathbf{C}^d$ to $W_x$, on which the unitary group $\mathrm{U}(d)$ of $\mathbf{C}^d$ acts freely and transitively, by means of

$$A \mapsto \left(e \mapsto e \circ A^{-1}\right). \tag{13.3}$$

It follows that

$$\mathrm{UF}\, W := \bigsqcup_{x \in M} \mathrm{UF}_x\, W$$

is a principal $\mathrm{U}(d)$-bundle over $M$, called the *unitary frame bundle of* $W$.

So we are in the situation of Theorem 13.1, with $G = \mathrm{U}(d)$, $\mathfrak{g} = \mathfrak{u}(d)$, the space of anti-selfadjoint $d \times d$-matrices, and $P = \mathrm{UF}\, W$. It is well-known that the ring of conjugacy-invariant polynomials on $\mathfrak{u}(d)$ is a generated by the elementary symmetric functions of the eigenvalues of $A \in \mathfrak{u}(d)$ without relations between the generators; or equivalently by the conjugacy-invariant, homogeneous polynomials $\tau_j(A)$ of degree $j$, defined by

$$\det_{\mathbf{C}}(1 + t\, A) = 1 + \sum_{j=1}^{d} t^j\, \tau_j(A). \tag{13.4}$$

Note that $\tau_1(A) = \mathrm{trace}_{\mathbf{C}}\, A$ and $\tau_d(A) = \det_{\mathbf{C}}\, A$.

There is a bijective correspondence between the $\mathrm{U}(d)$-invariant connections in $\mathrm{UF}\, W$ and the linear connections in $W$, which leave both the complex structure and the Hermitian structure invariant. This correspondence is defined by the condition that a local unitary frame $e$, a local section of $\mathrm{UF}\, W$, is horizontal at $x$, if and only if the vectorfields $e_j$ are covariantly constant at the point $x$. If $R^\nabla$ denotes the curvature operator of the covariant differentiation $\nabla$, defined by the linear connection in $W$, then (6.9) in this case reads:

$$-\,\Omega_{jk}\,(v_{\mathrm{hor}},\, w_{\mathrm{hor}}) = h_x\left(R_x^\nabla(v,\, w)\, e_k,\, e_j\right). \tag{13.5}$$

Here $v,\, w \in \mathrm{T}_x\, M$ and $1 \leq j,\, k \leq d$. That is, the matrix of $R^\nabla$ with respect to the basis $e_j$, of two-forms on $\mathrm{T}_x\, M$, is equal to minus the matrix of two-forms $\Omega \in \mathfrak{u}(d)$, pulled back by means of the horizontal lift $v \mapsto v_{\mathrm{hor}}$ from $\mathrm{T}_x\, M$ to $\mathrm{T}_{e_x}\, P$.

This leads to the following definition. The closed differential forms $\gamma_j \in \Omega^{2j}(M)$, $1 \leq j \leq d$, which occur in the decomposition

$$\det_{\mathbf{C}}\left(1 - \tfrac{1}{2\pi i}\Omega\right) = 1 + \sum_{j=1}^{d} \gamma_j \tag{13.6}$$

are called the *Chern forms* of the vector bundle $W$ with connection. The de Rham cohomology class $c_j := [\gamma_j] \in \mathrm{H}^{2j}(M)$, which does not depend on the choice of the connection in $W$, is called the *$j$-th Chern class of $W$*. Compare for instance Bott and Tu [BT, pp. 270, 267, 72, 73].

The factor $1/2\pi i$ has been put in, because it makes $c_j$ *integral*, that is, it belongs to the image in $\mathrm{H}^{2j}(M, \mathbf{C}) \simeq \mathrm{H}^{2j}(M)$, of the Čech cohomology group $\mathrm{H}^{2j}(M, \mathbf{Z})$ with integer coefficients, under the mapping from $\mathrm{H}^{2j}(M, \mathbf{Z})$ to $\mathrm{H}^{2j}(M, \mathbf{C})$ induced by the injection $\mathbf{Z} \to \mathbf{C}$. For complex line bundles, this will be analyzed in all details in Section 15.3.

The $j$-th Chern class has the topological interpretation that the integral of $c_j$ over a $2j$-dimensional cycle $C$ is equal to the intersection number of $C$ with the cycle $D$ of codimension $2j$, where generic $d - j + 1$ sections of $W$ are linearly dependent over $\mathbf{C}$. See Milnor and Stasheff [59, Th. on p. 306 and Def. on p. 158].

In order to simplify the notations, we will write

$$\Gamma = -\tfrac{1}{2\pi i}\Omega, \tag{13.7}$$

and call this $d \times d$-matrix of two-forms on $\mathrm{UF}\, W$ the *Chern matrix*, in view of (13.6). Note that in view of (13.5) the Chern matrix corresponds to $\frac{1}{2\pi i}$ times the curvature matrix.

We recall that if $f$ is an $\mathrm{Ad}\,\mathrm{U}(d)$-invariant polynomial on $\mathrm{u}(d)$ then, according to Theorem 13.1, $f(\Gamma)$ can be identified with a sum of closed differential forms of even degree in $M$. Actually, there is a unique polynomial $F$ in $d$ variables, such that, with the $\tau_j$ as in (13.4),

$$f(A) = F\left(\tau_1(A), \tau_2(A), \ldots, \tau_d(A)\right), \tag{13.8}$$

and we get

$$f(\Gamma) = F\left(\gamma_1, \gamma_2, \ldots, \gamma_d\right), \tag{13.9}$$

in which the closed form $\gamma_j \in \Omega^{2j}(M)$ is the $j$-th Chern form.

As a first application, if we replace, according to the remark after Theorem 11.1, the matrix of two-forms $R^L$ by $R^L/2\pi i$, which in view of (13.5) and (13.7) can be identified with the Chern matrix $\Gamma^L$ of the complex vector bundle $L$, then the last factor in (11.17) becomes equal to the characteristic form

$$\mathrm{ch}(L) := \mathrm{trace}_{\mathbf{C}}\left(\mathrm{e}^{\Gamma^L}\right) \tag{13.10}$$

which is called the *Chern character* of the vector bundle $L$ with connection. Actually, this name usually refers to the de Rham cohomology class of the right hand side in (13.10). Note that because of the finite-dimensionality of $M$, the Chern character of $L$ is a polynomial in the Chern classes of $L$ with rational coefficients. For instance, if $M$ is four-dimensional, then

$$\mathrm{ch}(L) = \dim L + c_1(L) + c_2(L) - \tfrac{1}{2}\, c_1(L)^2.$$

For the identification of the first factor in (11.17) with Chern forms, we are faced with the problem that the Levi-Civita connection in $\mathrm{T}\,M$ only leaves the almost complex structure $J$ invariant, if $(M,\, J,\, h)$ is a Kähler manifold, cf. the introduction of Chapter 3.

However, the first factor in (11.17) always has an interpretation in terms of characteristic forms for the principal $\mathrm{SO}(2n)$-bundle $\mathrm{SOF}\, M$ of oriented orthonormal frames in $\mathrm{T}\,M$. For any oriented Riemannian manifold $(M,\, \beta)$, the form

$$\alpha(M,\, \beta) := \left(\det \frac{R^\beta/2\pi i}{1 - \mathrm{e}^{-R^\beta/2\pi i}}\right)^{1/2} \tag{13.11}$$

is a sum of closed differential forms of even degree, and its de Rham cohomology class does not depend on the choice of the connection in $\mathrm{SOF}\, M$. If $M$ is compact, even-dimensional, and has a spin structure, then the integral over $M$ of the volume part of $\alpha$ is equal to the index of the Dirac operator acting on the spinor bundle, see the remark after Theorem 6.1. This is closely related to the *A-roof genus* of $M$, cf. Atiyah and Singer [7, Th. (5.3)].

In our case of an almost complex manifold, a translation in terms of Chern classes can be given in the following way. If we replace the covariant differentiation $\nabla^\beta$ in $\mathrm{T}\,M$, defined by the Levi-Civita connection, by a covariant differentiation $\tilde{\nabla}$ which leaves both $J$ and $\beta$ invariant, then we get another curvature operator $\tilde{R}^{\beta}$. Note that $\tilde{\nabla}$ is not torsion-free if $(M, J, h)$ is not a Kähler manifold. In view of (13.5) and (13.1), applied to the $\mathrm{SO}(2n)$-bundle SOF $M$, the resulting forms in (13.11) now differ by an exact one. Retricting oneself to unitary frames, we can apply Lemma 12.3, and express the right hand side of (12.13) in terms of Chern forms of T $M$, viewed as a complex vector bundle over $M$.

In terms of the Chern matrix

$$\Gamma^{\tilde{\nabla}} = -\tfrac{1}{2\pi i}\Omega^{\tilde{\nabla}},$$

in which $\Omega^{\tilde{\nabla}}$ denotes the curvature form of the connection in UF $M$ defined by $\tilde{\nabla}$, the sum

$$\mathrm{Td}(M, \tilde{\nabla}) := \det_{\mathbf{C}}\left(\frac{\Gamma^{\tilde{\nabla}}}{1 - \mathrm{e}^{-\Gamma^{\tilde{\nabla}}}}\right) \tag{13.12}$$

of closed differential forms of even degree is called the *Todd form* of the almost complex manifold $(M, J)$ with covariant differentiation $\tilde{\nabla}$. The de Rham cohomology class of (13.12) is called the *Todd class* $\mathrm{Td}(M, J)$ of the almost complex manifold $(M, J)$, it is independent of the choice of $\tilde{\nabla}$. It is a polynomial in the Chern classes of T $M$, viewed as a complex vector bundle over $M$, with rational coefficients. With these notations, we now have:

**Proposition 13.1** *The de Rham cohomology class of the volume part of (11.17) is equal to the part of degree $2n$ in*

$$\mathrm{Td}(M, J)\,\mathrm{ch}(L),$$

*the product of the Todd class of $(M, J)$ and the Chern character of L.*

*If $(M, J, h)$ is a Kähler manifold, then the volume part of (11.17) is equal to the product of the Todd form of $(M, J, h)$ and the Chern character form of L, defined by the respective connections in $\mathrm{T}M$ and L.*

## 13.3 The Lefschetz Formula

For the interpretation of (12.12) in terms of characteristic classes, there is the problem that in general the functions

$$X \mapsto \mathrm{trace}_{\mathbf{C}}\left(\gamma_{L,x}\, \mathrm{e}^{X}\right)$$

on $\mathfrak{u}(L_x)$ and

$$X \mapsto \det_{\mathbf{C}}\left(1 - {\gamma_{N,x}}^{-1}\, \mathrm{e}^{-X}\right)$$

on $\mathfrak{u}(N_x)$, respectively, are not conjugacy invariant. (Using unitary frames, we will conjugate these functions to functions on $\mathfrak{u}(\dim L)$ and $\mathfrak{u}(\dim N)$, respectively.) However, they are, if we restrict ourselves to the eigenspaces of $\gamma_{L,x}$ and $\gamma_{N,x}$ respectively. The following rigidity property ensures that this restriction leads to smooth objects on the fixed point manifolds of $\gamma$ in $M$.

**Lemma 13.1** *The linear transformations $\gamma_{L,x}$ and $\gamma_{N,x}$ in $L_x$ and $N_x$, respectively, are diagonalizable. The respective eigenvalues $\lambda_{L,j}$ and $\lambda_{N,k}$ are constant when $x$ varies in a connected component $F$ of $M^\gamma$, the fixed point set of $\gamma$ in $M$. The corresponding eigenspaces $L_{j,x} \subset L_x$ and $N_{k,x} \subset N_x$, with $x$ ranging over $F$, form smooth complex vector subbundles $L_j$ and $N_k$ of $L$ and $N$, respectively.*

**Proof** According to the remark in Section 10.1, we may assume that $\gamma$, or rather $\gamma_L$, belongs to a compact group $K$ of transformations. Now let $H$ be the closure in $K$ of the set of $\gamma^p$, with $p \in \mathbf{Z}$. Then $H$ is a compact abelian group of automorphisms of our structure, and $h(x) = x$ for each $x \in M^\gamma$, $h \in H$. We therefore have, for each $x \in M^\gamma$, the complex representations

$$h \mapsto \mathrm{T}_x\, h_M|_{N_x}$$

and

$$h \mapsto h_{L,x}$$

of $H$ in $N_x$ and $L_x$, respectively.

Let $\mathfrak{h}$ denote the Lie algebra of $H$ and $H^0 = \exp \mathfrak{h}$ the connected component of the identity element in $H$. Then $H^0$ is a torus and $H/H^0$ is isomorphic to $\mathbf{Z}/m\,\mathbf{Z}$, where $m$ is the smallest positive integer $p$ such that $\gamma^p \in H^0$. We write $\delta = \gamma^m$. Since $\delta \in H^0$, there exists an $X_0 \in \mathfrak{h}$ such that $\delta = \exp X_0$. It follows that

$$\tilde{\gamma} = \gamma \exp\left(-\tfrac{1}{m}X_0\right)$$

satisfies $\tilde{\gamma}^m = 1$, that is, $\tilde{\gamma}$ is periodic with period $m$.

In $\mathfrak{h}$ we have the lattice $\Lambda = \ker \exp$ of the $X \in \mathfrak{h}$, such that $\exp X = 1$; the torus $H^0$ is isomorphic to the additive group $\mathfrak{h}/\Lambda$. A complex-valued $\mathbf{R}$-linear form $\mu$ on $\mathfrak{h}$ is called a *weight* if

$$\mu(\Lambda) \subset 2\pi i\mathbf{Z}.$$

The weights form a lattice (= discrete additive subgroup) $W$ in $\mathfrak{h}^*$. For every linear representation $\rho$ of $H^0$ in a finite-dimensional complex vector space $V$ and $\mu \in W$, the weight space $V_\mu$ is defined as the set of $v \in V$, such that

$$\rho(\exp X)(v) = \mathrm{e}^{\langle X,\mu\rangle}\, v, \quad X \in \mathfrak{h}.$$

Then $V_\mu$ is a linear subspace of $V$, $V$ is equal to the direct sum of the $V_\mu$, $\mu \in W$, and there are only finitely many $\mu \in W$ such that $V_\mu \neq 0$, called the *weights of the representation* $\rho$.

Similarly, let $R_m$ denote the set of $z \in \mathbf{C}$, such that $z^m = 1$, the set of $m$-th roots of unity. If $\rho$ is a representation of $H$ in $V$, then for each $z \in R_m$, the space $V^z$ of the $v \in V$ such that $\rho(\tilde{\gamma})(v) = z\,v$ is a linear subspace of $V$ and we have that $V$ is equal to the direct sum of the subspaces

$$V_\mu^z := V_\mu \cap V^z, \quad \mu \in W,\ z \in R_m.$$

We now apply this to the representation of $H$ in the spaces $L_x$, $x \in F$; the representation in $N_x$ will be treated in an analogous way. The continuous dependence on $x \in F$, combined with the discreteness of the weight lattice

$W$ and the finite set $R_m$, yields that the complex vector bundle $L$ over $F$ has a direct sum decomposition into subbundles $L^j$, such that for each $j$ there are $\mu_j \in W$ and $z_j,\ z_k \in R_m$, with the property that

$$\exp X(l) = \mathrm{e}^{\langle X,\, \mu_j \rangle}\, l, \quad X \in \mathfrak{h},\ l \in L_{j,\, x}$$

and

$$\tilde{\gamma}(l) = z_j\, l, \quad l \in L_{j,\, x}.$$

It follows that the action of

$$\gamma = \tilde{\gamma}\ \exp\left(\tfrac{1}{m}\, X_0\right)$$

on $L_{j,\, x}$ is equal to multiplication by the complex number

$$\lambda_j = z_j\ \mathrm{e}^{\frac{1}{m}\, \langle X_0,\, \mu_j \rangle},$$

which does not depend on $x \in F$.

The constancy of the eigenvalues now also implies the smoothness of the eigenbundles $L_j$ and $N_k$. We give the proof for the $N_k$, the proof for the $L_j$ is analogous. Let $x_0 \in F$, and let $N_0 = N_{k,\, x_0}$ be the eigenspace in $N_{x_0}$ for the eigenvalue $\lambda_{N,\, k}$. Let $N_1$ be the sum of the $N_{k',\, x_0}$ with $k' \neq k$. Since $\gamma_N$ is a unitary transformation in $N_{x_0}$, it is diagonalizable, so $N_{x_0}$ is equal to the direct sum of $N_0$ and $N_1$. We write $v \in N_{x_0}$ as $v = v_0 + v_1$, with $v_0 \in N_0$ and $v_1 \in N_1$. In local coordinates in $F$ around $x_0$ and in a local trivialization of the vector bundle $N$, we now consider the linear inhomogeneous equation

$$\left(\mathrm{T}_x\, \gamma(v) - \lambda_{N,\, k}\, v\right)_1 = 0, \tag{13.13}$$

with $v_1$ as the unknown, and $x$ and $v_0$ as parameters. For $x = x_0$ and arbitrary $v_0$, $v_1 = 0$ is the unique solution, so there is an open neighborhood $U$ of $x_0$ in $F$, such that for each $x \in U$ and each $v_0$, there is a unique solution $v_1 = v_1(x,\, v_0)$, which depends linearly on $v_0$ and smoothly on $x$. In other words, (13.13) defines a smooth subbundle $\tilde{N}_k$ of $N$ over $U$. Moreover, if $x \in U$, then $N_{k,\, x} \subset \tilde{N}_{k,\, x}$, with equality if $x = x_0$.

For each $x \in F$, $N_x$ is equal to the direct sum over $k$ of the spaces $N_{k,\,x}$. Reading this for $x = x_0$, we get that $N_{x_0}$ is equal to the direct sum of the spaces $\tilde{N}_{k,\,x_0}$. By continuity, it follows that there is an open neighborhood $\tilde{U}$ of $x_0$ in $U$, such that for each $x \in \tilde{U}$, $N_x$ is equal to the direct sum over $k$ of the spaces $\tilde{N}_{k,\,x}$. Combining this with $N_{k,\,x} \subset \tilde{N}_{k,\,x}$ and the fact that $N_x$ is also equal to the direct sum over $k$ of the $N_{k,\,x}$, we arrive at $N_{k,\,x} = \tilde{N}_{k,\,x}$ for each $x \in \tilde{U}$. This completes the proof that, for each $k$, the $N_{k,\,x}$ with $x \in F$ form a smooth vector bundle over $F$. It is a complex vector bundle in the sense that it is $J$-invariant because $\mathrm{T}\,\gamma$ commutes with $J$. □

The curvature operators $R^L(v,\, w)$ leave the eigenbundles $L_j$ over $F$ invariant, where $L_j$ is the subbundle of $L$ where $\gamma_L$ acts as multiplication by the complex number $\lambda_{L,\,j}$. Moreover, by averaging over the compact group $H$ in the proof of Lemma 13.1, we can arrange that the covariant differentiation $\tilde{\nabla}$ of vector fields, which leaves the almost complex and the Riemannian structure invariant, also is invariant under $H$, hence under the action of $\mathrm{T}\,\gamma$ on $\mathrm{T}\,M$. As a consequence, the curvature operators $R^{\tilde{\nabla}}(v,\, w)$ leave the eigenbundles $N_k$ over $F$, on which $\gamma_N$ acts as multiplication by the complex numbers $\lambda_{N,\,k}$, invariant.

With regard to the action on $N$ of the curvature $R^{\beta}(v,\, w)$ of the Levi-Civita connection, we observe that if $\lambda_{N,\,k}$ is a non-real eigenvalue of $\gamma_N$, that is, if $\lambda_{N,\,k} \neq -1$, then the fact that $R^{\beta(v,\,w)}$ commutes with $\mathrm{T}\,\gamma$ implies that, in $N_k$, $R^{\beta}(v,\, w)$ commutes with $J$. So in this case the factor in the determinant of $1 - \gamma_N^{-1}\, \mathrm{e}^{R_N^{\beta}}$, which comes from $N_k$, can be identified with a characteristic form for the unitary group. That is, it can be expressed in terms of the Chern forms of the complex vector bundle $N_k$ over $F$. On the other hand, if $\lambda_{N,\,k} = -1$, then we get a characteristic form for the full rotation group of the corresponding eigenspace.

Applying the arguments for Proposition 13.1, with $W$ in Theorem 13.1 replaced by $L_j$ and $N_k$, we now get:

**Proposition 13.2** *Let $F$ be a connected component of $M^\gamma$, the fixed point set of $\gamma$ in $F$. Let $\lambda_{L,j}$ and $\lambda_{N,k}$ be the eigenvalues of the action of $\gamma$ on L and $N$, respectively and let $L_j$ and $N_k$ be the corresponding eigenbundles.*

*Then, for $x \in F$, the de Rham cohomology class of the $F$-volume part of (12.12) is equal to the part of degree $2n_F$, of the element*

$$\mathrm{Td}(F,\, J) \sum_j \lambda_{L,j}\, \mathrm{ch}(L_j) / \left[ \prod_k \det_{\mathbf{C}} \left( 1 - \lambda_{N,k}{}^{-1}\, \mathrm{e}^{-\tilde{\Gamma}(N_k)} \right) \right]$$

*of $\mathrm{H}^*_{\text{de Rham}}(F)$. Here $\tilde{\Gamma}(N_k)$ denotes the Chern matrix of the bundle $N_k$ over $F$, for a connection in $TM$ which leaves both $J$ and $\beta$ invariant, but which need not be torsion-free.*

*The identity holds on the level of characteristic forms, if $(F,\, J,\, h)$ is a Kähler manifold and, for each $x \in F$, the curvature operators $R^\beta_x(v,\, w)$ of the Levi-Civita connection commute with $J_x$ in the eigenspace of $T_x\,\gamma$ for the eigenvalue $-1$.*

The formula in Proposition 13.2 agrees with the one of Patodi [65], and of Atiyah and Bott [5, Th. 4.12] in the case of isolated fixed points. In [5, Sec. 5], one can also find the application to torus actions on complex flag manifolds, in which case the formula reduces to the formula of Hermann Weyl for the characters of the irreducible representations of compact connected Lie groups. In the comparison one has to keep in mind, that in [5] and [65] the action of the automorphism on differential forms is the pullback, whereas we take the inverse of the pullback in order to get a homomorphism from the group of automorphisms to the group of linear transformations in the space of differential forms.[1]

[1] I am grateful to Yael Karshon for pointing this out to me; it led to a correction of a previous sign mistake which I had made in the exponent of $\lambda_{N,k}$. In Atiyah and Singer [7, Th. (4.6)], it seems that the action of $\gamma$ on differential forms is by means of the inverse of the pullback.

## 13.4 A Simple Example

The following simple case can be used as a test for the sign of the Chern matrix in the exponential appearing in the denominator. Let $M = \mathrm{P}(2, \mathbf{C})$ be the complex projective plane. With the Fubini-Study metric, the unique $\mathrm{PU}(3)$-invariant one, $M$ is a Kähler manifold. We consider the action on $M$ of $\lambda \in \mathbf{C}$, $\lambda \neq 1$, $|\lambda| = 1$, given in projective coordinates by

$$\gamma = \gamma_\lambda : [z_1, z_2, z_3] \mapsto [\lambda z_1, z_2, z_3].$$

The fixed point set has two connected components $F_0$, $F_1$. $F_0$ is the isolated point $z_2 = z_3 = 0$; $\gamma$ acts on the tangent space as multiplication by $1/\lambda$. The other connected component $F_1$ is the complex projective line $z_1 = 0$; here $\gamma$ acts on the normal bundle as multiplication by $\lambda$.

Using that $\mathrm{H}^j(F_1) = 0$ for $j > 2$, we can write the Todd class of $F_1$ as $1 + t$, with $t \in \mathrm{H}^2(F_1)$. Let $c \in \mathrm{H}^2(F_1)$ denote the Chern class of the normal bundle of $F_1$ in $M$. According to Proposition 13.2, the alternating sum of the traces of the action of $\gamma$, on the cohomology groups $\mathrm{H}^q(M, \mathcal{O})$, is equal to

$$\chi = \chi(\lambda) = (1 - 1/\lambda)^{-2} + \int_{F_1} \mathrm{area}\left(\frac{1+t}{1 - \lambda\, \mathrm{e}^{-c}}\right).$$

Using that $c^2 = 0$ on $F_1$ we get

$$\left(1 - \lambda\, \mathrm{e}^{-c}\right)^{-1} = (1-\lambda)^{-1} \left(1 + (1-\lambda)^{-1} \lambda\, c\right)^{-1} = (1-\lambda)^{-1} - (1-\lambda)^{-2} \lambda\, c.$$

It follows that

$$\chi(\lambda) = (1-\lambda)^{-2} \left(\lambda^2 - (T + C)\, \lambda + T\right),$$

in which

$$T = \int_{F_1} t, \quad C = \int_{F_1} c.$$

By its definition, $\chi(\lambda)$ depends in a polynomial fashion on $\lambda$, so we have necessarily that $\chi(\lambda) \equiv 1$, or $T + C = 2$ and $T = 1$, or $T = C = 1$.

Note that if the contribution from the isolated fixed point is as stated, then replacing $\lambda$ by $\lambda^{-1}$ in the contribution from $F_1$ would not lead to a polynomial $\chi(\lambda)$. And that replacing $e^{-c}$ by $e^{c}$ in the contribution from $F_1$ would amount to changing $C$ to $-C$.

In order to check the formula, we now use that, for any $n$ and $q > 0$,

$$\mathrm{H}^q(\mathrm{P}(n, \mathbf{C}), \mathcal{O}) = 0,$$

whereas

$$\mathrm{H}^0(\mathrm{P}(n, \mathbf{C}), \mathcal{O}) = \mathbf{C},$$

corresponding with the constant functions on $\mathrm{P}(n, \mathbf{C})$. See for instance Griffiths and Harris, [33, p. 49]. This shows that indeed $\chi(\lambda) \equiv 1$. In view of the Riemann-Roch formula for $F_1$ this also implies that $T = 1$. That $C = 1$ agrees with the interpretation of the Chern class of $F_1$ as the zero locus of a generic section of the normal bundle, counted with orientation. Indeed, any (nearby) complex projective line intersects $F_1$ in exactly one point and with positive orientation.

# Chapter 14

# The Orbifold Version

The local formula is particularly suited for the generalization of the Lefschetz formula to compact orbifolds. The result is stated in Theorem 14.1 at the end of this chapter. In its formulation, we need certain auxiliary orbifolds, which we call the fixed point orbifolds and which will be introduced in Section 14.4. We begin this chapter with the definition of orbifolds and then discuss how the spin-c Dirac operator and the corresponding heat kernel can be introduced on these.

## 14.1 Orbifolds

Loosely speaking a $d$-dimensional *orbifold* is a space which locally can be identified with the quotient of $\mathbf{R}^d$ by the action of a finite group of orthogonal linear transformations.

More precisely, an orbifold is a paracompact Hausdorff topological space $M$, provided with a collection $\mathcal{C}$ of so-called *orbifold charts*. Each such chart is a quadruple $(U, V, H, p)$, in which $U$ is an open subset of $M$, $V$ is a smooth $d$-dimensional manifold, $H$ is a finite group acting on $V$, and $p : V \to U$ is a continuous mapping which induces a homeomorphism from the orbit space $V/H$ onto $U$. It is required that the $U$ cover $M$. Moreover, if $(U_i, V_i, H_i, p_i) \in \mathcal{C}$, $i = 1, 2$ and $x \in U_1 \cap U_2$, then there exists $(U, V, H, p) \in \mathcal{C}$ such that $x \in U \subset U_1 \cap U_2$. To the orbifold structure also belongs that in this situation, there are injective group homomorphisms $\rho_i : H \to H_i$ and $\rho_i$-equivariant open embeddings $\phi_i : V \to V_i$ such that

$$\rho_i(H) = \{h_i \in H_i \mid h_i \cdot \phi_i(V) \cap \phi_i(V) \neq \emptyset\}.$$

This property implies that $p_i \circ \phi_i = p$ and that the stabilizer subgroups (also called isotropy subgroups) $H_v$ in $H$ of points $v$ over a given point $x \in M$ are isomorphic. Finally it is required that for each $x \in M$ there exists $(U, V, H, p) \in \mathcal{C}$ and $v \in V$, such that $p(v) = x$ and $H = H_v$. The latter is equivalent with the condition that $p^{-1}(\{x\}) = \{v\}$.

If a compact Lie group $H$ acts smoothly on a manifold $V$ with a fixed point $v$, then there exists a smooth coordinate system around $v$ in $V$ in which the action consists of linear transformations, cf. Bochner [12, Th. 1]. (Using an $H$-invariant Riemannian structure in $V$, this can be seen as a consequence of (10.6).) With an orbifold chart as in the last requirement in the definition of orbifolds, we find a coordinate neighborhood in $V$ in which the action of $H_v$ is linear, so locally $M$ can be identified with the quotient of $\mathbf{R}^d$ by the action of a finite group of linear transformations. By averaging an arbitray inner product over $H_v$ we get an invariant inner product, so one can also arrange that the transformations are orthogonal.

The definition of orbifold given here differs slightly from the definition of a *V-manifold* of Satake [66], [67], in that it is allowed that elements $h \in H$ act as reflections in hyperplanes and that it is not assumed that the $H$ act effectively on the manifolds $V$. Although orbifolds in general are not smooth, one can do differential geometry in them in almost the same way as in smooth manifolds. For instance, a mapping $\pi$ from an orbifold $\tilde{M}$ to an orbifold $M$ is called *smooth*, if in orbifolds charts $(\tilde{U}, \tilde{V}, \tilde{H}, \tilde{p})$ and $(U, V, H, p)$ of $\tilde{M}$ and $M$, respectively, one has a smooth mapping $\phi : \tilde{V} \to V$ and a homomorphism $\rho : \tilde{H} \to H$ such that $\phi$ is $\rho$-equivariant and $p \circ \phi = \pi \circ \tilde{p}$ in $\tilde{U}$.

An *orbifold vector bundle* over $M$ consists of an orbifold $L$ and a smooth mapping $\pi : L \to M$ with the additional requirement that corresponding to the orbifold chart $(U,\ V,\ H,\ p)$ of $M$ we have an orbifold chart $(\tilde{U},\ \tilde{V},\ \tilde{H},\ \tilde{p})$ of $L = \tilde{M}$ as above, with the additional condition that $\phi : \tilde{V} \to V$ is a smooth vector bundle and $H = \tilde{H}$ acts on $\tilde{V}$ by means of vector bundle automorphisms. It should be noted here that the fiber $\pi^{-1}(\{x\})$ in $L$ over a point $x = p(v) \in M$ can be identified with the quotient of the vector space $\tilde{V}_v = \phi^{-1}(\{v\})$ by means of the isotropy group $H_v$, so in general the fiber need not even be homeomorphic to a vector space. (For reasons like this, when proving statements about orbifolds one usually works in the smooth local branched coverings rather than in the space itself.) A smooth section of the orbifold vector bundle is a smooth mapping $s : M \to L$ such that $\pi \circ s$ is equal to the identity in $M$. It is characterized by the property that the corresponding smooth mappings $\sigma : V \to \tilde{V}$ are smooth $H$-equivariant sections of the vector bundle $\phi : \tilde{V} \to V$.

As an example, smooth differential forms in $M$ correspond to smooth $H$-invariant differential forms in the local branched coverings $V$, which can be identified with smooth $H$-equivariant sections of the exterior bundles $\Lambda^p \mathrm{T}^* V$. In this way, smooth differential forms in $M$ are sections of an orbifold vector bundle $\Lambda^p \mathrm{T}^* M$ over $M$. The point of Satake [66], [67] is that with these definitions, the de Rham theory of smooth manifolds can be generalized to orbifolds, practically without any change.

For a locally free action of a compact Lie group $G$ on a smooth manifold $N$, the orbit space $M = N/G$ has a natural orbifold structure. Indeed, for each $y \in N$ the stabilizer subgroup $G_y$ of $y$ in $G$ is finite. The local linearization of the $G_y$-action leads to the existence of a so-called *slice* at $y$ for the action of $G$, which is a smooth $G_x$-invariant manifold $V$ through $y$, such that $\mathrm{T}_y N = \mathrm{T}_y V \oplus \mathrm{T}_y(G \cdot y)$ and such that each nearby $G$-orbit in $N$ intersects $V$ in a $G_y$-orbit in $V$. In this way neighborhoods $U$ in $N/G$ of orbits $x = G \cdot y$ are identified with quotients $V/G_y$ of smooth manifolds by finite groups $G_y$, which yield the desired orbifold charts.

It is a remark of Kawasaki [44, p. 76], that conversely every orbifold $M$ can be obtained in this way, by providing $M$ with a smooth Riemannian structure (in the orbifold sense) and then taking for $N$ the orthonormal frame bundle $N = \mathrm{OF}\, M$ (in the orbifold sense). Then $N$ is a smooth manifold on which $\mathrm{O}(d)$ acts locally freely, and $M \simeq N/\, \mathrm{O}(d)$. We will not use this construction in the sequel, but instead keep working with the finite groups $H$ acting in the local manifolds $V$.

For any action of a Lie group $G$ on a manifold $N$, one says that $y,\ z \in N$ belong to the same *orbit type* if there exists $h \in G$, such that

$$G_z = h \cdot G_y \cdot h^{-1} \quad (= G_{h \cdot y})\,.$$

This means that the orbit $G \cdot z$ is diffeomorphic to the orbit $G \cdot y = G \cdot (h \cdot y)$ via the natural identifications of $G/G_z = G/G_{h \cdot y}$ with $G \cdot z$ and $G \cdot y$ by means of the mappings $g \mapsto g \cdot z$ and $g \mapsto g \cdot h \cdot y$, respectively. The *orbit types in* $N$, the equivalence classes for this equivalence relation, are $G$-invariant subsets of $N$. Their quotients by $G$, viewed as subsets of $N/G$, are called the *orbit types in* $N/G$.

Now suppose that the action of $G$ is proper, which implies that all stabilizer subgroups are compact. Using slices in which the action of the stabilizer groups are linear, one can verify that the connected components $C$ of the orbit types in $N/G$ are quotients $\tilde{C}/G$ of smooth submanifolds $\tilde{C}$ of $N$. Moreover, the condition that $\tilde{C}$ has only one orbit type yields that $C = \tilde{C}/G$ can be provided with the structure of a smooth manifold in a natural way. The orbit types in $N$ form the strata of a Whitney stratification of $N$. This is a locally finite decomposition of $N$ into smooth submanifolds with certain properties about how each stratum behaves near its boundary; where the latter is a locally finite union of lower dimensional strata. We will call this stratification of $N$ the *orbit type stratification of* $N$. The corresponding locally finite decomposition of $N/G$ into smooth manifolds will be called the *orbit type stratification of* $N/G$.

If we apply this to the action of $H$ in $V$, for an orbifold chart $(U,\ V,\ H,\ p)$, then we get an orbit type stratification of $V$ and one of the open subset $U \simeq V/H$ of the orbifold $M$. The compatibility conditions for the orbifold charts give that the orbit type stratifications of $U_1$ and $U_2$ agree in an orbifold neighborhood $U$ in $U_1 \cap U_2$ of a point in $U_1 \cap U_2$. Piecing together the local strata in $M$ we get a locally finite decomposition of $M$ into connected smooth submanifolds, which is called the *orbifold stratification* of $M$. The way how the strata behave near their boundaries is studied best in the local charts in $V$ in which the stabilizer groups act by means of linear transformations. It also follows from the compatibility conditions that if $(U_i, V_i, H_i, p_i) \in \mathcal{C}$, $i = 1, 2$ are orbifold charts of $M$ and $x_i \in U_i$ belong to the same stratum $S$ of $M$, then the stabilizer subgroups $(H_i)_{x_i}$ are isomorphic, so in particular they have the same number $m$ of elements. The number $m = m(S)$ will be called the *multiplicity* of the stratum $S$ of $M$.

It is known that if $M$ is connected, then there is only one stratum $S$ for which $m(S)$ is minimal. This stratum, called the *principal stratum*, is open and dense in $M$ and all other strata have strictly lower dimension. The multiplicity of the principal stratum is called the *multiplicity of the orbifold* $M$, and denoted by $m(M)$. We have $m(M) = 1$ if and only if the actions of the groups $H$ are effective. We also note that for an orbifold bundle $L$ over $M$, it can happen that $m(M) > m(L)$, in which case $L$ is called an *improper bundle* over $M$. This happens for instance with the normal bundle in $M$ of a lower dimensional stratum.

If $\rho$ is a smooth density with compact support in $M$, then it is natural to define the integral of $\rho$ over $M$ as the integral over the principal stratum $S$ of the smooth density $\rho|_S$ in the smooth manifold $S$. If $(U, V, H, p)$ is an orbifold chart and $\operatorname{supp}\rho \subset U$, then the integral of $\rho_V$ over $V$ is equal to $\#(H/H_V)$ times the integral of $\rho$ over $U$, where $H_V$ is the normal subgroup of $H$ consisting of the $h \in H$ which act as the identity on $V$. Because $\#(H_V) = m(M)$, we get

$$\frac{1}{m(M)} \int_M \rho = \frac{1}{\#(H)} \int_V \rho_V \quad \text{if } \operatorname{supp}\rho \subset U. \tag{14.1}$$

## 14.2 The Virtual Character

In order to define the orbifold version of the virtual character (1.1), we assume now that $M$ is compact as a topological space and has a smooth almost complex structure and Hermitian structure. That is, the dimension $d$ is even, say $2n$, and the $V$ in the orbifold charts are provided with $H$-invariant smooth almost complex and Hermitian structures. Furthermore it is assumed that $L$ is an orbifold complex vector bundle over $M$ with a Hermitian connection, which means that the $L_V = \tilde{V}$ are complex vector bundles over the $V$ with Hermitian connection, for which the finite groups $H$ preserve the complex and Hermitian structures of the fibers and the Hermitian connection in the bundle. The duals of the canonical line bundles of the $V$ piece together to an orbifold complex line bundle over $M$ which will be denoted by $K^*$, and we assume that it too is provided with a Hermitian connection. It may be remarked that, given the almost complex structure, all the other structures can be introduced by averaging arbitrary such structures in the manifolds $V$ and $\tilde{V}$ over the finite groups $H$, and then piecing these together by means of a smooth partition of unity over $M$, subordinate to the covering by the sets $U$.

Let $E_V^+$ and $E_V^-$ be the direct sum over the even and odd $q$, respectively, of the bundles of $(0, q)$-forms over $V$. These piece together to orbifold complex vector bundles $E^+$ and $E^-$ over $M$. The spin-c Dirac operator $D_V^\pm$ of $V$ acts on the space $\Gamma_V^\pm$ of smooth sections of the bundle $E_V^\pm \otimes L_V$ over $V$ and it commutes with the action of $H$. It therefore maps invariant sections to invariant sections; in this way the operators $D_V^\pm$ piece together to an operator $D^\pm$ which maps the space $\Gamma^\pm$ of smooth sections (in the orbifold sense) of $E^\pm \otimes L$ to $\Gamma^\mp$. The operator $D = D^+ \oplus D^-$ is called the *orbifold spin-c Dirac operator* of $M$.

The compactness of $M$ together with the ellipticity of $D^\pm$ (that is, of the operators $D_V^\pm$) yield again that the kernels $N^\pm$ of the linear operators $D^\pm$ are finite-dimensional complex vector spaces. If $\gamma$ is a bundle automorphism of

$L$ (in the orbifold sense) which leaves all the given structures invariant, then it induces an operator $\gamma^{\pm}$ in $\Gamma^{\pm}$ such that $D^{\pm} \circ \gamma^{\pm} = \gamma^{\mp} \circ D^{\pm}$. In particular $\gamma^{\pm}$ leaves the space $N^{\pm}$ invariant, which allows us to define the the virtual character $\chi(\gamma)$ as in (1.1).

## 14.3 The Heat Kernel Method

The next step is to compute the virtual character by means of the heat kernel method, as explained in Chapter 7. For the construction of the integral kernel $K(t, x, y)$ of the heat diffusion operator $T(t)$ of (7.7), we begin with an asymptotic heat kernel $K_V(t, x, y)$ for $t \downarrow 0$, defined on $\mathbf{R} \times V \times V$, for the operator $D_V$ on the smooth local branched covering $V$ in an orbifold chart $(U, V, H, p)$. For the existence and uniqueness of asymptotic heat kernels, see Theorem 8.1, here we use that it is not required that the manifold is compact, because of the local nature of the expansion. Because $Q_V = {D_V}^2$ commutes with the $H$-action, the integral operator $T_V(t)$ defined by a smooth invariant representation of the asymptotic kernel $K_V$ commutes with the $H$-action as well. In particular it maps $H$-invariant sections with compact support in $V$ to $H$-invariant sections over $V$ and thus defines a linear operator $T_U(t)$ which maps section of $\Gamma$ with compact support in $U$ to sections of $\Gamma$ over $U$.

Let $\phi_j$ be a smooth partition of unity in $M$, with the property that supp $\phi_j \cup$ supp $\phi_k$ is contained in a $U = U(j, k)$ of an orbifold chart, when supp $\phi_j \cap$ supp $\phi_k \neq \emptyset$; let $I$ be the set of $(j, k)$ for which the latter condition holds. Write

$$\tilde{T}(t) = \sum_{(j, k) \in I} \phi_j \, T_{U(j, k)} \, \phi_k .$$

Then $\tilde{T}(t)$ is an asymptotic solution of the heat equation (7.8) in the sense that $d\tilde{T}(t)/dt + Q \circ \tilde{T}(t)$ is an integral operator with a smooth kernel which for any $N$ is of order $\mathcal{O}\left(t^N\right)$ as $t \downarrow 0$. Moreover, $\tilde{T}(t) \to I$ as $t \downarrow 0$. As in

[9, Section 2.3-5], one gets a unique solution $T(t)$ of (7.8) on the compact orbifold $M$ which is an integral operator with smooth kernel, for which in addition $\tilde{T}(t)$ is an approximation in the sense that, for any $N$ and in the topology of inetrgal operators with smooth kernels,

$$T(t) = \tilde{T}(t) + \mathcal{O}\left(t^N\right) \quad \text{as } t \downarrow 0.$$

Here an operator $T$ is called an integral operator with a smooth kernel acting on sections of an orbifold vector bundle $L$ over an orbifold $M$ if it has the following property. If $(U, V, H, p)$ and $(U', V', H', p')$ are orbifold charts of $M$, then there is an integral operator $T_{V', V}$ with smooth kernel $K_{V', V}$ defined on $V' \times V$, such that for each section $u$ with compact support in $U$, for which $u_V$ is the corresponding $H$-invariant section over $V$, $T\, u$ is given in $U'$ as the $h'$-invariant section $T_{V', V} u$ over $V'$. Because the $U'$'s cover $M$, this determines $T\, u$ and using a partition of unity subordinate to the $U$'s we see that $T$ is completely determined by the $K_{V', V}$.

Integral operators with smooth kernels on compact orbifolds are of trace class and one has the same generalities about their trace as in the case of smooth compact manifolds. Let the $\phi_j$ be a partition of unity as above. If $A$ and $B$ are operators such that $A$ is of trace class and $B$ is bounded, then $A\, B$ and $B\, A$ are of trace class and $\text{trace}_{\mathbf{c}}(A\, B) = \text{trace}_{\mathbf{c}}(B\, A)$. Using this we get that

$$\begin{aligned} \text{trace}_{\mathbf{c}}\, T &= \sum_{j,\, k} \text{trace}_{\mathbf{c}}\, (\phi_j\, T\, \phi_k) = \sum_{j,\, k} \text{trace}_{\mathbf{c}}\, (T\, \phi_k\, \phi_j) \\ &= \sum_{(j,\, k) \in I} \text{trace}_{\mathbf{c}}\, (T\, \phi_k\, \phi_j) = \sum_{(j,\, k) \in I} \text{trace}_{\mathbf{c}}\, (\phi_j\, T\, \phi_k)\, , \end{aligned}$$

because $\phi_k\, \phi_j = 0$ when $\text{supp}\, \phi_j \cap \text{supp}\, \phi_k = \emptyset$. Now $\phi_j\, T\, \phi_k$ is an integral operator with a smooth kernel which has compact support in $U \times U$, where $U = U(j,\, k)$. If the $\psi_r$ are the $H$-invariant smooth functions in $V = V(j,\, k)$ which define $\phi_r$, then $\phi_j\, T\, \phi_k$ is represented by the integral operator $T_{j,\, k}$ in $V$ with kernel equal to $K_{j,\, k} = \psi_j\, K_{V,\, V}\, \psi_k$, with $V = V(j,\, k)$.

We now have to compute the trace of $T_{j,\,k}$, but as an operator on the space $\Gamma^H$ of $H$-invariant sections instead of on the space $\Gamma$ of all sections, so we cannot use the formula (7.2) right away. However, it is a basic principle of the theory of representations of compact groups, that the orthogonal projection $\Pi$ from $\Gamma$ to $\Gamma^H$ is given by averaging the elements of $\Gamma$ over $H$. Note that if $A : \Gamma \to \Gamma$ is of trace class, then

$$\text{trace}_{\mathbf{C}}\,(\Pi\, A) = \text{trace}_{\mathbf{C}}\,(A\,\Pi) = \text{trace}_{\mathbf{C}}\left(A\,\Pi^2\right) = \text{trace}_{\mathbf{C}}\,(\Pi\, A\,\Pi)\,,$$

where the latter is equal to the trace of the restriction of $A$ to $\Gamma^H$. So we get that

$$\begin{aligned}\text{trace}_{\mathbf{C}}\,(\phi_j\, T\, \phi_k) &= \frac{1}{\#(H)} \sum_{h\in H} \text{trace}_{\mathbf{C}}\,(h \circ T_{j,\,k}) \\ &= \frac{1}{\#(H)} \sum_{h\in H} \text{trace}_{\mathbf{C}}\,(h\, \psi_j\, T_{V,\,V}\, \psi_k)\,, \end{aligned} \tag{14.2}$$

where $V = V(j,\, k)$.

In our application to the spin-c Dirac operator we take $T = \gamma \circ T(t)$, so $h \circ T_{V,\,V} = h \circ \gamma_V \circ T(t)_{V,\,V}$, replace the trace by the supertrace, and replace the integral by its limit for $t \downarrow 0$. Each integral in the right hand side of (14.2) then becomes equal to the integral over the fixed point set $V^{h\circ\gamma_V}$ of $h \circ \gamma_V$ in $V$ of a density which is equal to $\psi_k\, \psi_j$ times the volume part of (12.12); *with* $\gamma$ *replaced by* $h \circ \gamma_V$. Summing over the $(j,\, k)$, where the restricition $(j,\, k) \in I$ can be dropped because $\psi_k\, \psi_j = 0$ if $(j,\, k) \notin I$, we get rid of the factors $\psi_k\, \psi_j$.

## 14.4 The Fixed Point Orbifolds

Our next task is to pass from the local descriptions of the integration in the orbifold charts to invariantly defined integrals over suitable orbifolds $\tilde{F}$, which we expect to lie in the fixed point set $M^\gamma$ of $\gamma$ in $M$, because $y \in V$ is a fixed point of $h \circ \gamma_V$ for some $h \in H$, if and only $x = p(y)$ is a fixed point of

$\gamma$. If $x \in M^\gamma$ then we may choose an orbifold chart $(U, V, H, p)$ such that $p^{-1}\{x\} = \{v\}$ and $H = H_v$; it follows that $\gamma_V(v) = v$. The compatibility condition for $\gamma_V$ means that there exists an automorphism $\alpha$ of $H$ such that, for each $a \in H$, $\gamma_V \circ a \circ {\gamma_V}^{-1} = \alpha(a)$.

For the construction of the orbifolds $\tilde{F}$, which unfortunately in general are not quite suborbifolds of $M$, we start by considering the space

$$\hat{V} = \{(v,\, h) \in V \times H \mid (h \circ \gamma_V)\,(v) = v\},$$

which is the disjoint union of the manifolds $V^{h \circ \gamma_V} \times \{h\}$, $h \in H$. On $\hat{V}$, each $a \in H$ acts by means of the transformation

$$(h,\, v) \mapsto \left(a(h),\, a \cdot h \cdot \alpha(a)^{-1}\right).$$

Indeed, if $(h \circ \gamma_V)(v) = v$ then $\alpha(a)^{-1} \circ \gamma_V = \gamma_V \circ a^{-1} \circ {\gamma_V}^{-1} \circ \gamma_V = \gamma_V \circ a^{-1}$ yields that

$$\left(a \cdot h \cdot \alpha(a)^{-1}\right) \circ \gamma_V \circ a(v) = a \circ h \circ \gamma_V(v) = a(v).$$

For a given dimension $\delta$ of a fixed point set, let $\hat{V}(\delta)$ be the union of the connected components of $\hat{V}$ which have dimension equal to $\delta$. Then the compatibility conditions allow us to piece together the $\hat{V}(\delta)/H$ to a $\delta$-dimensional orbifold $\tilde{M}^\gamma(\delta)$. We write $\tilde{M}^\gamma$ for the disjoint and disconnected union over $\delta$ of the $\tilde{M}^\gamma(\delta)$. That is, $\tilde{M}^\gamma$ is the disjoint and disconnected union of orbifolds which is obtained by piecing together the spaces $\hat{V}/H$. The connected components $\tilde{F}$ of $\tilde{M}^\gamma$ will be called the *fixed point orbifolds* for the automorphism $\gamma$ of $M$.

For each connected component $\tilde{F}$ of $\tilde{M}^\gamma$ the projection

$$\hat{V} \ni (v,\, h) \mapsto p(v) \in M$$

induces a mapping $\tau : \tilde{F} \to M$, which actually is an immersion of orbifolds. Because $\tau$ is proper (with finite fibers) and the image is closed in $M$, we also get that each orbifold $\tilde{F}$ is compact. The image $\tau(\tilde{F})$ is contained in some connected component $F$ of $M^\gamma$. For a given connected component $F$ of $M^\gamma$ there is one distinguished connected component $F'$ of $\tilde{M}^\gamma$, which corresponds to taking $h = 1$ in each orbifold chart; in this case $\tau$ is an embedding from $F'$ into $M$.

## 14.5 The Normal Eigenbundles

At a fixed point $v$ of $h \circ \gamma_V$, $h \circ \gamma_V$ acts as a unitary transformation on $\mathrm{T}_v V$, so we get a decomposition into eigenspaces, where the eigenvalues are complex numbers on the unit circle and the eigenspace for the eigenvalue 1 is equal to the tangent space at $v$ of the fixed point manifold of $h \circ \gamma_V$. The eigenspaces are complex vector spaces, so have even real dimension. Note that for $\gamma = 1$ this yields that the real dimension of each orbifold stratum of $M$ is even. In particular Satake's condition, that the $h \in H$ are no reflections in hyperplanes, is satisfied.

Because the group of isometries with a given fixed point is compact, we can find local coordinates around $v$ in $V$ in which $\gamma_V$ and all the $h \in H$ act as linear transformations. This yields that the eigenvalues of the tangent action of $h \circ \gamma_V$ are constant along the fixed point set of $h \circ \gamma_V$. Moreover in these coordinates the eigenspaces are constant as well. As in Lemma 13.1, we therefore get finitely many complex numbers $\lambda_{\tilde{N},\,k}$, on the unit circle and not equal to 1, and a decomposition of the normal bundle $\tilde{N}$ of $\tilde{F}$ in $M$ into orbifold subbundles $\tilde{N}_k$, such that in a $V$ the fiber of $\tilde{N}_k$ is equal to the nonzero eigenspace of $h \circ \gamma_V$ with the eigenvalue $\lambda_{\tilde{N},\,k}$.

More can be said if $\gamma$ belongs to a one-parameter group $t \mapsto \mathrm{e}^{t\,X}$ of automorphisms of $M$. Here $X$ acts as a smooth vector field $X_M$ on $M$ in the orbifold sense, which in $V$ is represented by a smooth vector field $X_V$, with flow equal to $\gamma_V = \mathrm{e}^{t\,X_V}$. Because the automorphism group of a finite group is finite, the one-parameter group of conjugations with $\mathrm{e}^{t\,X_V}$ reduces to the identity, so $\gamma_V$ commutes with the $h \in H$. Or, equivalently, $X_V$ is

$H$-invariant.

Let $E_j$ be the eigenspace of the linearization of the vector field $X_V$, with eigenvalue $2\pi i \mu_j$, $\mu_j \in \mathbf{R}$. Then $\gamma_V$ acts on $E_j$ as multiplication by $\mathrm{e}^{2\pi i t \mu_j}$. $E_j$ splits further into the eigenspaces $E_{j,j'}$ of $h$ for the eigenvalues $\epsilon_{j'}$, where the latter are roots of untity, because $h^N = 1$ for some positive integer $N$. It follows that for given $\lambda$, the eigenspace of $h \circ \gamma_V$ for the eigenvalue $\lambda$ is equal to the sum of the $E_{j,j'}$ such that

$$\epsilon_{j'} \, \mathrm{e}^{2\pi i t \mu_j} = \lambda.$$

We will say that $\gamma = \mathrm{e}^{tX}$ is *at resonance with the orbifold structure* if it happens that more than one $E_{j,j'}$ is involved. The set of $t$ for which this occurs is a discrete subset of $\mathbf{R}$, so generically we have that only one $E_{j,j'}$ occurs for every eigenvalue $\lambda$.

In the nonresonant case we get in particular that $\tilde{F}$ in $V$ is represented by the intersection of the zeroset of $X_V$ and the fixed point set of $h$, where in turn the zeroset of $X$ is equal to $M^{\gamma}$. The distinguished orbifold $F'$, corresponding to taking $h = 1$, is mapped by $\tau$ onto a connected component $F$ of $M^{\gamma}$, so $F$ is a suborbifold of $M$ and $F'$ is identified with $F$ by means of the diffeomorphism $\tau : F' \to F$.

Moreover, every orbit type $S_V$ for the action of $H$ in $V^{\gamma_V}$ is equal to a connected component of the complement $C$ of finitely many complex linear subspaces of $V^{h \circ \gamma_V} = V^h \cap V^{\gamma_V}$, for some $h \in H$. However, $C$ itself is already connected; the conclusion is that if $\tau(\tilde{F}) \subset F$, then $\tau(\tilde{F})$ is equal to the closure of a stratum $S$ of the orbifold $F$, and for each stratum $S$ of $F$ there exists at least one fixed point orbifold $\tilde{F}$ for which $\tau(\tilde{F}) = \overline{S}$. So in the nonresonant case the closures of the strata of $F$ are immersed suborbifolds of $F$. For a given stratum $S$, there may exist more than one fixed point orbifold $\tilde{F}$ such that $\tau(\tilde{F}) = \overline{S}$.

A similar decomposition can be obtained of the pullback $\tilde{L}$ of $L$ to $\tilde{F}$, by means of the immersion $\tau : \tilde{F} \to M$. Here the subbundles are denoted by

$\tilde{L}_j$, and the corresponding complex numbers $\lambda_{\tilde{L},j}$ on the unit circles are the eigenvalues of the unitary linear actions of the $h \circ \gamma_V$ on the fibers of $L_V$.

## 14.6 The Lefschetz Formula

As in Theorem 12.1, we get that the short time limit of the supertrace of $\gamma$ on the kernel of the spin-c Dirac operator is equal to the sum over the fixed point orbifolds $\tilde{F}$ of the integral over $\tilde{F}$ of $1/m(\tilde{F})$ times the volume part of (12.12), where the latter is defined in the regular branched coverings $V$ in terms of the smooth transformation $h \circ \gamma_V$ and the smooth connections of the various bundles. The factor $1/m(\tilde{F})$ appears because we had the factor $1/\#(H)$ from the averaging over $H$ and we have the formula (14.1) for integration over orbifolds.

The connection in the tangent bundle of $M$ which is used in the definition of the spin-c Dirac operator is the Levi-Civita connection. According to the warning after (3.2), the Levi-Civita connection leaves the almost complex structure invariant if and only if the orbifold $M$ is Kähler, in which case we have the simplification (12.14). In the non-Kähler case we introduce another connection in $\mathrm{T}\, M$ which leaves both the Riemannian and the almost complex structure invariant. This can be done by averaging an arbitrary such connection in $V$ over $H$ and piecing together the resulting connections in the $U$ by means of a partition of unity.

It follows from the proof of Weil's theorem in Section 16.4 and the proof of Theorem 16.1, that the difference of (12.14) and the second line in (12.12) is equal to the exterior derivative of a form $\gamma_V$, where $\gamma_V$ is defined pointwise in terms of the aformentioned connection in the normal eigenbundle of $\gamma^{h \circ \gamma_V}$. In particular the $\gamma_V$ are $H$-invariant, so they piece together to a smooth differential form in $\tilde{F}$ in the orbifold sense. Using the orbifold version of Stokes' theorem, we get that the integral over $\tilde{F}$ of the volume part of (12.12) does not change if we replace the second line in

(12.12) by (12.14). However, as explained in Section 13.3, the latter can be expressed in terms of the Chern matrices $\Gamma\left(\tilde{N}_k\right)$ of the normal eigenbundles $\tilde{N}_k$, as in Proposition 13.2.

In this way we arrive at the following orbifold version of the Lefschetz fixed point theorem.

**Theorem 14.1** *Let $M$ be a compact orbifold, provided with an almost complex and Hermitian structure, and a Hermitian connection in the dual $K^*$ of the canonical line bundle of $M$. Let $L$ be a complex orbifold vector bundle over $M$ with a Hermitian connection and let $D^{\pm} : \Gamma^{\pm} \longrightarrow \Gamma^{\mp}$ be the corresponding spin-c Dirac operator. Write $N^{\pm}$ for the kernel of $D^{\pm}$, which is a finite-dimensional complex vector space.*

*Second, let $\gamma$ be an automorphism of all structures involved. Let $\mathcal{O}^{\gamma}$ denote the collection of fixed point orbifolds $\tilde{F}$ as defined in Section 14.4. As explained in Section 14.5, we have the splitting of $\tilde{L} = \tau^* L$ into eigenbundles $\tilde{L}_j$ over $\tilde{F}$ with eigenvalues $\lambda_{\tilde{L},j}$, and the splitting of the normal bundle $\tilde{N}$ of $\tilde{F}$ in $M$ into the eigenbundles $\tilde{N}_k$, with eigenvalues $\lambda_{\tilde{N},k}$. Then we have*

$$\mathrm{trace}_{\mathbf{C}}\left(\gamma|_{N^+}\right) - \mathrm{trace}_{\mathbf{C}}\left(\gamma|_{N^-}\right) = \sum_{\tilde{F}\in\mathcal{O}^{\gamma}} \frac{1}{m(\tilde{F})} \int_{\tilde{F}} \alpha_{\tilde{F}} \tag{14.3}$$

*in which $\alpha_{\tilde{F}}$ is equal to the part of degree* $\dim\tilde{F}$ *of the characteristic class*

$$\mathrm{Td}(\tilde{F},\, J) \sum_j \lambda_{\tilde{L},j}\, \mathrm{ch}(\tilde{L}_j) / \left[\prod_k \det_{\mathbf{C}}\left(1 - \lambda_{\tilde{N},k}{}^{-1}\, \mathrm{e}^{-\tilde{\Gamma}(\tilde{N}_k)}\right)\right] \tag{14.4}$$

*in* $\mathrm{H}^*_{\text{de Rham}}(\tilde{F})$. *Here* $\mathrm{ch}(\tilde{L}_j)$ *and* $\Gamma\left(\tilde{N}_k\right)$ *denote the Chern character of $\tilde{L}_j$ and the Chern matrix of $\tilde{N}_k$, respectively.*

**Remark 1** If the almost complex structure is integrable, which means that $M$ is a complex analytic orbifold, then the left hand side of (14.3) is equal to the orbifold holomorphic Lefschetz number

$$\chi(\gamma) = \chi_{M,L}(\gamma) := \sum_{q=0}^{n} (-1)^q\, \mathrm{trace}_{\mathbf{C}}\, \gamma|_{\mathrm{H}^q(M,\mathcal{O}(L))}. \tag{14.5}$$

The proof is as in Section 2.

If $\gamma = 1$ (in which case there is no resonance), then the left hand side of (14.3) is equal to the index of $D^+$, because $D^-$ is equal to the adjoint of $D^+$. In (14.4) the eigenvalues $\lambda_{\tilde{L},j}$ and $\lambda_{\tilde{N},k}$ then are roots of unity which come from the orbifold structure. The contribution from the distinguished $\tilde{F} = M$ is equal to $1/m(M)$ times the integral over $M$ of the product of the Todd class of $M$ and the Chern character of $L$. However, even if $m(M) = 1$, there are also contributions from the lower dimensional strata of $M$, if $M$ is not a smooth manifold.

If $\gamma = 1$ and $M$ is a complex analytic orbifold, then the left hand side of (14.3) is equal to the Riemann-Roch number (2.20).

**Remark 2** If $\gamma$ belongs to a one-parameter group $t \mapsto \mathrm{e}^{tX}$ of automorphisms, then there exist real numbers $\mu_{\tilde{L},j}$ and $\mu_{\tilde{N},k}$, and roots of unity $\epsilon_{\tilde{L},j}$ and $\epsilon_{\tilde{N},k}$ such that, for $t$ in the complement of a discrete subset of $\mathbf{R}$, there is no resonance of $\gamma$ with the orbifold structure and

$$\lambda_{\tilde{L},j} = \epsilon_{\tilde{L},j}\ \mathrm{e}^{2\pi i t \mu_{\tilde{L},j}} \tag{14.6}$$

and

$$\lambda_{\tilde{N},k} = \epsilon_{\tilde{N},k}\ \mathrm{e}^{2\pi i t \mu_{\tilde{N},k}}\,. \tag{14.7}$$

In this case $M^\gamma$ is equal to the set of zeros of $X_M$ in $M$.

Moreover, the sum over the $\tilde{F}$ in (14.3) can be organized as the sum over the connected components $F$ of $M^\gamma$ of the sum of the orbifold strata $S$ of $F$ of the sum of the fixed point orbifolds $\tilde{F}$ such that $\tau(\tilde{F}) = \overline{S}$. The integral over $\tilde{F}$ then can be viewed as an integral over $S$. For each $F$ there is one distinguished summand for which $\tilde{F} = \overline{S} = F$, where $S$ is the principal stratum of $F$.

**Remark 3** The virtual character does not depend on the choice of the Riemannian structure in $M$ and the Hermitian structure and connection in $L$ and $K^*$ and moreover does not change under small perturbations of the almost complex structure in $M$. It depends on the linearization of the action of $\gamma$ on the normal bundles in $M$ of the fixed point orbifolds, but apart from this geometric information it can be viewed as a purely topological number. This is especially true for the index of $D^+$, which is a topological number in terms of the orbifold bundle $L$ and the almost complex structure in the orbifold $M$. Here the linearizations of the actions of the groups $H_v$, especially their eigenvalues, are considered to be a part of the topological structure of the orbifold bundle $L$ and the orbifold $M$.

# Chapter 15

# Application to Symplectic Geometry

In this chapter we will describe how, starting from a manifold $M$ with a symplectic form $\sigma$, which satisfies an integrality condition such that it is the Chern form of a connection of a suitable complex line bundle $L$ over $M$, one can get all the structures needed for the definition of a spin-c Dirac operator $D$, acting on sections of $E \otimes L$. Furthermore a Hamiltonian action of a torus $T$ in $(M, \sigma)$ can be lifted to $L$ in such a way that it preserves all the previously introduced structures. If $M$ is compact, then one obtains a fixed point formula for the virtual character of the representation in the finite-dimensional null space of $D$. The ingredients in this formula which come from the topology of the line bundle $L$, and the $T$-action on $L$ at the fixed points of the $T$-action in $M$, are expressed in terms of the symplectic form $\sigma$ and the momentum mup $\mu$ of the Hamiltonian action, respectively. For the reason that we want to include the case that $(M, \sigma)$ itself is a reduced phase space, which usually is an orbifold rather than a smooth manifold, we actually will apply the orbifold version of the fixed point formula of Chapter 14. The detailed results are presented in Section 15.5, which the reader may consult first in order to get a quick impression.

In the sections preceding it we give background material on symplectic manifolds, Hamiltonian group actions and reduction, the complex line bundle $L$ and the lifting of the action to $L$. The text of these sections is not very narrowly aimed at our final goal, in that they proceed from generalities to the more special situation considered in Section 15.5. Also we mention a number of facts which may be of interest to the reader but which are not strictly needed in Section 15.5.

## 15.1 Symplectic Manifolds

A background reference for this section is Abraham and Marsden [1, Ch. 3].

Let $M$ be a smooth manifold. A *symplectic structure* on $M$ is a smooth exterior differential form $\sigma$ of degree two on $M$ which is closed, that is $\mathrm{d}\,\sigma = 0$, and which has the property that, for every $m \in M$, the antisymmetric bilinear form $\sigma_m$ on $\mathrm{T}_m\, M$ is nondegenerate, which means that if $v \in \mathrm{T}_m\, M$ and $\sigma_m(v,\, w) = 0$ for every $w \in \mathrm{T}_m\, M$, then $v = 0$.

The last condition implies that $M$ has even dimension, say equal to $2n$. More precisely, the *Darboux lemma* says that for every $m \in M$ there exists a system of local coordinates $q_1, \ldots, q_n, p_1, \ldots, p_n$ in an open neighborhood $U$ of $m$ in $M$, such that

$$\sigma = \sum_{j=1}^{n} \mathrm{d}\, p_j \wedge \mathrm{d}\, q_j \quad \text{in } U. \tag{15.1}$$

The condition that $\mathrm{d}\,\sigma = 0$ can be viewed as a flatness condition, which yields that locally all symplectic structures in manifolds of the same dimension are isomorphic.

One of the prime examples of a symplectic manifold is the cotangent bundle $\mathrm{T}^*\, Q$ of an arbitrary manifold $Q$, with $\sigma$ equal to the exterior derivative of the tautological one-form $\tau$ in $\mathrm{T}^*\, Q$, cf. (8.29), (8.30). In classical mechanics $\mathrm{T}^*\, Q$ is called the *phase space* corresponding to the configuration space $Q$.

The flow of a vector field $v$ in $M$ leaves $\sigma$ invariant if and only if

$$0 = \mathcal{L}_v\,\sigma = \mathrm{i}(v)\,(\mathrm{d}\sigma) + \mathrm{d}\,(\mathrm{i}(v)\,\sigma) = \mathrm{d}\,(\mathrm{i}(v)\,\sigma)$$

which locally is equivalent to

$$\mathrm{i}(v)\,\sigma = -\,\mathrm{d}f \tag{15.2}$$

for some smooth function $f$, determined up to addition of a constant. The nondegeneracy of $\sigma$ implies that for each smooth function $f$ on $M$ there is a unique vector field $v$ in $M$ which satisfies (15.2), $v$ is called the *Hamilton vector field* defined by the function $f$ and is denoted by $\mathrm{H}_f$. The minus sign in (15.2) is used because in Darboux coordinates the differential equations $dm/dt = \mathrm{H}_f(m)$ for the flow of $\mathrm{H}_f$ take the familiar form

$$\left.\begin{array}{rcl} \frac{dq_j}{dt} & = & \frac{\partial f(q,p)}{\partial p_j} \\ \frac{dp_j}{dt} & = & -\frac{\partial f(q,p)}{\partial q_j} \end{array}\right., \quad 1 \leq j \leq n, \tag{15.3}$$

of a Hamilton system in classical mechanics.

If $f$ and $g$ are smooth functions on $M$, then their *Poisson brackets* $\{f,\, g\}$ is the smooth function on $M$ defined by

$$\{f,\, g\} = \langle \mathrm{H}_f,\, \mathrm{d}g\rangle = \sigma\,(\mathrm{H}_f,\, \mathrm{H}_g)\,. \tag{15.4}$$

The right hand side shows that the Poisson brackets are anti-symmetric, that is, $\{g,\, f\} = -\{f,\, g\}$. In local Darboux coordinates, the Poisson brackets can also be written as

$$\{f,\, g\} = \sum_{j=1}^{n} \frac{\partial f}{\partial p_j}\,\frac{\partial g}{\partial q_j} - \frac{\partial f}{\partial q_j}\,\frac{\partial g}{\partial p_j}.$$

The invariance of the symplectic form under the $\mathrm{H}_f$-flow $\Phi_t$ yields, if $\delta$ denotes differentiation with respect to $t$ at $t = 0$, that

$$[\mathrm{H}_f,\, \mathrm{H}_g] = \delta\Phi_t^*\,\mathrm{H}_g = \delta\,\mathrm{H}_{\Phi_t^* g} = \mathrm{H}_{\{f,g\}}\,.$$

In turn this implies that

$$\{f, \{g, h\}\} + \{g, \{h, g\}\} + \{h, \{f, g\}\} = (\mathrm{H}_f \circ \mathrm{H}_g - \mathrm{H}_g \circ \mathrm{H}_f)\, h - \mathrm{H}_{\{f, g\}}\, h = 0.$$

In other words:

*With the Poisson brackets, the space $C^\infty(M)$ of smooth functions on $M$ is a Lie algebra, and $f \mapsto H_f$ is a homomorphism of Lie algebras from $C^\infty(M)$ to the Lie algebra $\mathcal{X}^\infty(M)$ of smooth vector fields in $M$.*

By definition, the image consists of the Lie algebra of Hamilton vector fields, whereas the kernel consists of the space $\mathrm{H}^0_{\text{de Rham}}(M)$ of functions which are constant on the connected components of $M$.

We have already seen the example (8.28) of the kinetic energy as a function of the momenta, for which the $\mathrm{H}_f$-solution curves project to the geodesics in $Q$ of the corresponding Riemannian structure in $Q$. An application of the fact that the flow of a Hamilton vector field leaves $\sigma$ invariant is that it also leaves the canonical volume form $\omega = \pm\frac{1}{n!}\sigma^n$, equal to the volume part of $\pm\,\mathrm{e}^\sigma$, invariant. Here we have chosen $\pm = (-1)^{n(n+1)/2}$, in order to get that in Darboux coordinates we have

$$\omega = \mathrm{d}q_1 \wedge \ldots \wedge \mathrm{d}q_1 \wedge \mathrm{d}p_1 \wedge \ldots \wedge \mathrm{d}p_n.$$

In classical mechanics the invariance of the canonical volume in $M$ under Hamiltonian flows is usually referred to as Liouville's theorem.

A (locally defined) diffeomorphism $\Phi$ in a manifold $Q$ induces a diffeomorphism $\tilde{\Phi}$ in $\mathrm{T}^*\, Q$, which for each $q \in Q$ is defined on $\mathrm{T}^*_q\, Q$ as the inverse of the transposed mapping of $\mathrm{T}_q\, Q$. If $v$ is a vector field in a manifold $Q$, let $\tilde{v}$ be the function on $\mathrm{T}^*\, Q$ which is defined by

$$\tilde{v}(p) = \langle v_q, p \rangle \quad \text{if } q \in Q,\ p \in (\mathrm{T}_q\, Q)^*. \tag{15.5}$$

Then the flow of the Hamilton vector field $\mathrm{H}_{\tilde{v}}$ in $\mathrm{T}^*\, Q$ is equal to flow in $\mathrm{T}^*\, Q$ which is induced by the flow of $v$ in $Q$. In formula:

$$\mathrm{e}^{t\, \mathrm{H}_{\tilde{v}}} = \widetilde{\mathrm{e}^{t\, v}}, \quad t \in \mathbf{R}.$$

A special example occurs when $Q$ is equal to a Lie group $G$. Then each $h \in G$ acts on $G$ by means of left multiplication

$$\mathrm{L}(h) : g \mapsto h\,g : G \to G$$

and by right multiplication

$$\mathrm{R}(h) : g \mapsto g\,h : G \to G.$$

Let Diff$(Q)$ denote the group of diffeomorphisms of a manifold $Q$. Then the associativity of the multiplication in $G$ is equivalent to each of the following statements:

i) $\mathrm{L} : h \mapsto \mathrm{L}(h)$ is a homomorphism from $G$ to Diff$(G)$.

ii) $\mathrm{R}' : h \mapsto \mathrm{R}(h)^{-1} = \mathrm{R}\left(h^{-1}\right)$ is a homomorphism from $G$ to Diff$(G)$.

iii) For each $a,\, b \in G$ we have $\mathrm{L}(a) \circ \mathrm{L}(b) = \mathrm{L}(b) \circ \mathrm{L}(a)$.

That is, left and right multiplication induce two commuting group actions L and R$'$ of $G$ on itself.

The action

$$h \mapsto \mathrm{Ad}\, h := \mathrm{L}(h) \circ \mathrm{R}(h)^{-1}$$

of conjugation by $h$ is called the *adjoint* action. Its tangent map at the identity element 1 is an automorphism of the Lie algebra $\mathfrak{g} = \mathrm{T}_1\, G$ of $G$, which also is denoted by Ad $h$. The homomorphism

$$\mathrm{Ad} : h \mapsto \mathrm{Ad}\, h : G \to \mathrm{Aut}(\mathfrak{g})$$

is called the *adjoint representation* of $G$ in $\mathfrak{g}$.

For each $X \in \mathfrak{g}$, the infinitesimal left action

$$\mathrm{L}(X) := \tfrac{d}{dt}\, \mathrm{L}\left(\mathrm{e}^{t\,X}\right)|_{t=0}$$

is the vector field on $G$ which at the point $g \in G$ is given by

$$\mathrm{L}(X)_g = \mathrm{T}_1\, \mathrm{R}(g)(X).$$

It follows that the corresponding Hamilton function $\widetilde{\mathrm{L}(X)}$ on $\mathrm{T}^* G$ is given by

$$\widetilde{\mathrm{L}(X)}(p) = \langle X, \, \mathrm{T}_1 \, \mathrm{R}(g)^* p \rangle \quad \text{if } p \in \mathrm{T}_g^* \, G. \tag{15.6}$$

In other words, $\widetilde{\mathrm{L}(X)}$ is the right invariant extension to $\mathrm{T}^* G$ of the linear function $X : \xi \mapsto \langle X, \, \xi \rangle$ on the dual space $\mathfrak{g}^*$ of the Lie algebra $\mathfrak{g}$. If we use left multiplications to translate back to the fiber $\mathfrak{g}^*$ of $\mathrm{T}^* G$ over the identity element, then we get

$$\widetilde{\mathrm{L}(X)} \left( \widetilde{\mathrm{L}(g)}(\xi) \right) = \langle X, \, \mathrm{Ad}^* \, g(\xi) \rangle \quad g \in G, \; \xi \in \mathfrak{g}^*. \tag{15.7}$$

Here

$$\mathrm{Ad}^* : h \mapsto \mathrm{Ad}^* \, h := \left( \mathrm{Ad} \, h^{-1} \right)^* = \left( (\mathrm{Ad} \, h)^* \right)^{-1} \tag{15.8}$$

is the so-called *coadjoint representation* of $G$ in $\mathfrak{g}^*$.

Similarly, the infinitesimal right action $\mathrm{R}'(X) = -\mathrm{R}(X)$ induces the Hamilton function on $\mathrm{T}^* G$ given by

$$\widetilde{\mathrm{R}'(X)}(p) = -\langle X, \, \mathrm{T}_1 \, \mathrm{L}(g)^* p \rangle \quad \text{if } p \in \mathrm{T}_g^* \, G, \tag{15.9}$$

which is the left invariant extension of the linear function $-X$ on $\mathfrak{g}^*$. It is also determined by

$$\widetilde{\mathrm{R}'(X)} \left( \widetilde{\mathrm{R}'(g)}(\xi) \right) = -\langle X, \, \mathrm{Ad}^* \, g(\xi) \rangle \quad g \in G, \; \xi \in \mathfrak{g}^*. \tag{15.10}$$

## 15.2 Hamiltonian Group Actions and Reduction

A background reference for this section is [1, Ch. 4], or the review [20].

Let $G$ be a connected Lie group which acts on $M$. One says that the action of $G$ in the symplectic manifold $(M, \, \sigma)$ is *Hamiltonian* with *momentum mapping* $\mu$, if the following two conditions hold.

i) $\mu$ is a smooth mapping from $M$ to $\mathfrak{g}^*$ which intertwines the action of $G$ in $M$ with the coadjoint action of $G$ in $\mathfrak{g}^*$.

ii) For each $X \in \mathfrak{g}$ the infinitesimal action $X_M$ of $X$ in $M$ is equal to the Hamilton vector field $\mathrm{H}_{\mu_X}$ of the smooth function

$$\mu_X : m \mapsto \langle X,\, \mu(m)\rangle \tag{15.11}$$

on $M$. In equation:

$$\mathrm{i}\,(X_M)\,\sigma + \mathrm{d}\langle X,\, \mu\rangle = 0,\ X \in \mathfrak{g}. \tag{15.12}$$

Note that any other mapping $\tilde{\mu} : M \to \mathfrak{g}^*$ is a momentum mapping for the Hamiltonian action of $G$, if and only if there exists a constant element $\nu$ in $[\mathfrak{g},\, \mathfrak{g}]^0$ such that $\tilde{\mu} = \mu + \nu$. Here

$$\begin{aligned} [\mathfrak{g},\, \mathfrak{g}]^0 &:= \{\xi \in \mathfrak{g}^* \mid \langle [X,\, Y],\, \xi\rangle = 0\ \ \forall X,\, Y \in \mathfrak{g}\} \\ &= \{\xi \in \mathfrak{g}^* \mid \mathrm{Ad}^*\, g(\xi) = \xi\ \ \forall g \in G\} \end{aligned} \tag{15.13}$$

is the set of fixed points in $\mathfrak{g}^*$ for the coadjoint action. We say in this case that $\tilde{\mu}$ is obtained from $\mu$ by means of a *neutral shift*. The freedom of changing the momentum mapping by means of a neutral shift will be used in Proposition 15.4 about Hamiltonian torus actions, in which case $[\mathfrak{g},\, \mathfrak{g}] = 0$, so we can shift over any constant element of $\mathfrak{g}^*$. On the other hand, if $[\mathfrak{g},\, \mathfrak{g}] = \mathfrak{g}$, which for instance happens when $\mathfrak{g}$ is semisimple, then there is no freedom in the choice of the momentum mapping at all.

Since the flow of any Hamilton vector field leaves $\sigma$ invariant, we get that each $g \in \exp \mathfrak{g}$ leaves $\sigma$ invariant. And since the connected Lie group $G$ is generated by $\exp \mathfrak{g}$, it follows that the action of $G$ in $M$ leaves $\sigma$ invariant. In general the condition for the action to be Hamiltonian is somewhat stronger than to leave $\sigma$ invariant.

If $G$ acts on a manifold $Q$ then we have, for each $q \in Q$, the linear mapping $\alpha_q : \mathfrak{g} \to \mathrm{T}_q\, Q$, defined by

$$\alpha_q(X) = \tfrac{d}{dt}\, \mathrm{e}^{t\,X}(q)|_{t=0}, \quad X \in \mathfrak{g}.$$

If $\mu : \mathrm{T}^* Q \longrightarrow \mathfrak{g}^*$ is the mapping which on $\mathrm{T}^*_q Q$ is equal to the transposed of $\alpha_q$, then we see from (15.5) that the induced action of $G$ on $\mathrm{T}^* Q$ is Hamiltonian with $\mu$ as its momentum mapping. This is one of the main sources of Hamiltonian group actions in classical mechanics, where $G$ is viewed as a symmetry group of the configuration space.

Returning to the Hamiltonian action in a general symplectic manifold $M$, let $f$ be a smooth function on $M$ which is invariant under the action of $G$. Then we get for each $X \in \mathfrak{g}$ that

$$0 = \mathrm{H}_{\langle X,\,\mu\rangle} f = \{\langle X,\, \mu\rangle,\, f\} = -\{f,\, \langle X,\, \mu\rangle\} = -\,\mathrm{H}_f \langle X,\, \mu\rangle,$$

from which we see that every coordinate $\langle X,\, \mu\rangle$ of the momentum is constant along the $\mathrm{H}_f$-solution curves. In clasical mechanics, the conclusion that for every Hamiltonian symmetry group of the Hamiltonian function the momenta are constants of motion for the Hamiltonian system, is called *Noether's principle.* Put in another way, it says that for every $\xi \in \mathfrak{g}^*$ the level set

$$\mu^{-1}(\{\xi\}) = \{m \in M \mid \mu(m) = \xi\},$$

the fiber of $\xi$ of the mapping $\mu : M \longrightarrow \mathfrak{g}^*$, is invariant under the flow of $\mathrm{H}_f$.

Disregarding $f$ and its Hamiltonian flow, we take a closer look at the level set $\mu^{-1}(\{\xi\})$. If $\xi$ is a regular value of $\mu$, which means that for every $m \in \mu^{-1}(\{\xi\})$ the linear mapping $\mathrm{T}_m\, \mu : \mathrm{T}_m\, M \longrightarrow \mathfrak{g}^*$ is surjective, then $\mu^{-1}(\{\xi\})$ is a smooth submanifold of $M$, of codimension equal to the dimension of $G$. Condition i) for Hamiltonian actions implies that $g$ maps $\mu^{-1}(\{\xi\})$ to $\mu^{-1}(\{\mathrm{Ad}^*\, g(\xi)\})$. This implies that on $\mu^{-1}(\{\xi\})$ we still have the action of the stabilizer subgroup

$$G_\xi := \{g \in G \mid \mathrm{Ad}^*\, g(\xi) = \xi$$

of $\xi$ in $G$, for the coadjoint action.

Let us write $X_M = \mathrm{H}_{\langle X,\,\mu\rangle}$ for the infinitesimal action of $X \in \mathfrak{g}$ in $M$. Then we have, for every $v \in \mathrm{T}_m\, M$,

$$\sigma_m\,(v,\, X_{M,m}) = \langle v,\, \mathrm{d}\langle X,\, \mu\rangle_m\rangle = \langle X,\, \mathrm{T}_m\, \mu(v)\rangle. \tag{15.14}$$

It follows that $X_{M,m} = 0$ if and only $X$ is orthogonal to the image of $\mathrm{T}_m\,\mu$, which by the assumption that $\mu(m) = \xi$ is a regular value means that $X = 0$. That is:

*The action of $G$ is locally free at all points of $\mu^{-1}(\{\xi\})$.*

It also follows from (15.14) that $v$ is tangent to $\mu^{-1}(\{\xi\})$ if and only if $v$ is in the symplectic orthogonal complement of all $X_{M,m}$, $X \in \mathfrak{g}$. In other words:

*The tangent space at $m$ of $\mu^{-1}(\{\xi\})$ is equal to the symplectic orthogonal complement of the tangent space $T_m(G \cdot m)$ at $m$ of the $G$-orbit through the point $m$, defined as the set of all $X_{M,m}$ such that $X \in \mathfrak{g}$.*

On the other hand property i) of the Hamiltonian action implies that, for each $X \in \mathfrak{g}$,

$$\mathrm{T}_m\,\mu\,(X_{M,m}) = -(\mathrm{ad}\,X)^*\,(\mu(m))\,.$$

The inner product of the left hand side with $Y \in \mathfrak{g}$ is equal to the inner product of $X_{M,m}$ with the derivative at of the function $\langle Y,\, \mu\rangle$, so it is equal to the value at $m$ of $\sigma\,(X_M,\, Y_M)$. In other words, we have that

$$\sigma\,(X_M,\, Y_M) = -\langle [X,\, Y]\,,\, \mu\rangle,\;\; X,\, Y \in \mathfrak{g}. \tag{15.15}$$

From this we see:

*In $T_m\,\mu^{-1}(\{\xi\}) = T_m(G \cdot m)$ the kernel of the symplectic form is equal to the set of $X_{M,m}$ such that $X \in \mathfrak{g}_\xi$, which is the tangent space to the $G_\xi$-orbit through $m$.*

If we strengthen the property that the action of $G_\xi$ on the level set $\mu^{-1}(\{\xi\})$ is locally free to the condition that it is *proper and free*, then its $G_\xi$-orbit space

$$M_\xi = \mu^{-1}(\{\xi\})/G_\xi \tag{15.16}$$

is a smooth manifold, of dimension equal to $\dim M - \dim G - \dim G_\xi$. Let $\iota_\xi$ denote the identity embedding $\mu^{-1}(\{\xi\}) \to M$ and $\pi_\xi$ the projection

$\mu^{-1}(\{\xi\}) \to M_\xi$ which assigns to each $m \in \mu^{-1}(\{\xi\})$ its $G_\xi$-orbit. As a consequence of the fact that the tangent spaces of the $G_\xi$-orbits are equal to the null spaces of the restriction of the symplectic form $\sigma$ to the level set $\mu^{-1}(\{\xi\})$, we get:

There is a unique two-form form $\sigma_\xi$ in $M_\xi$, such that $\iota_\xi^* \sigma = \pi_\xi^* \sigma_\xi$. Moreover, $\sigma_\xi$ is a symplectic form in $M_\xi$.

Here the second statement follows because

$$\pi_\xi^* \, \mathrm{d}\sigma_\xi = \mathrm{d}\, \pi_\xi^* \sigma_\xi = \mathrm{d}\, \iota_\xi^* \sigma = \iota_\xi^* \, \mathrm{d}\sigma = 0,$$

hence $\mathrm{d}\sigma_\xi = 0$ because $\pi_\xi$ is a submersion. The symplectic manifold $(M_\xi, \sigma_\xi)$ is called the *Marsden-Weinstein reduced phase space at the level* $\xi$, after it was introduced in [56].

**Remark** Suppose $f$ is a smooth function on $M$ such that

$$\forall X \in \mathfrak{g} \quad X_M f = 0 \text{ on } \mu^{-1}(\{\xi\}). \tag{15.17}$$

This condition is satisfied if $f$ is $G$-invariant and in turn it implies that the restriciton of $f$ to $\mu^{-1}(\{\xi\})$ is $G_\xi$-invariant. Let $f_\xi$ be the unique smooth function on $M_\xi$, such that $\iota_\xi^* f = \pi_\xi^* f_\xi$. The condition (15.17 means that $\mathrm{H}_f$ is tangent to $\mu^{-1}(\{\xi\})$, and we get that $\pi_\xi$ intertwines the restriction of $\mathrm{H}_f$ to $\mu^{-1}(\{\xi\})$ with $H_{f_\xi}$, the Hamilton vector field of $f_\xi$ in the reduced phase space $M_\xi$. That is, if $\phi^t$ and $\phi_\xi^t$ are the flows in $\mu^{-1}(\{\xi\})$ and $M_\xi$ of the vector fields $\mathrm{H}_f$ and $\mathrm{H}_{f_\xi}$, respectively, then $\pi_\xi \circ \phi^t = \phi_\xi^t \circ \pi_\xi$.

In many situations, for instance if the group $G_\xi$ is compact (in which case $\xi$ is sometimes called an elliptic element), the action of $G_\xi$ on the level set $\mu^{-1}(\{\xi\})$ is proper. Although there are many interesting examples where the action is also free, there are at least as many cases where this condition does not hold. In these cases, as discussed in Section 14.1, the reduced phase space is an *orbifold* rather than a smooth manifold. That is, a wealth of interesting examples of symplectic orbifolds arise as reduced phase spaces for Hamiltonian group actions.

An interesting example of a Hamiltonian action is the action $\tilde{L}$ of $G$ on its own cotangent bundle $\mathrm{T}^* G$, which is induced by the left multiplications in $G$. In this case the restriction of the momentum mapping to $\mathfrak{g}^* = \mathrm{T}_1^* G \subset \mathrm{T}^* G$ is equal to the identity in $\mathfrak{g}^*$, so every $\xi \in \mathfrak{g}^*$ is a regular value of $\mu_{\tilde{L}}$. Because the action $L$ of $G$ on itself by left multiplications is proper and free, an the action $\tilde{L}$ covers $L$, we conclude that the action $\tilde{L}$ is proper and free as well; we conclude that the reduced phase space is a smooth symplectic manifold.

The level set $N_\xi = \mu_{\tilde{L}}^{-1}(\{\xi\})$ is equal to the orbit $\tilde{R}'(G)\xi$ of $\xi$ for the action $\tilde{R}'$ of $G$ in $\mathrm{T}^* G$, which is induced by the right multiplications in $G$. Because also $\tilde{R}'$ is proper and free, this leads to a canonical identification of the reduced phase space $N_\xi / G_\xi$ with the homogeneous space $G/G_\xi$. The symplectic form $\sigma_\xi$ of the reduced phase space thereby corresponds to the unique $G$-invariant two-form on $G/G_\xi$ which on the tangent space $\mathfrak{g}/\mathfrak{g}_\xi$ of $G/G_\xi$ at $1 \cdot G_\xi$ is given by

$$\sigma_\xi(X, Y) = -\langle [X, Y], \xi \rangle, \quad X, Y \in \mathfrak{g}/\mathfrak{g}_\xi. \tag{15.18}$$

See (15.15).

On the other hand one has the natural identification of $G/G_\xi$ with the coadjoint orbit $\mathcal{O}_\xi = \mathrm{Ad}^* G(\xi)$ of $\xi$ in $\mathfrak{g}^*$, where the identification is induced by the mapping $g \mapsto \mathrm{Ad}^* g(\xi)$ from $G$ to $\mathfrak{g}^*$. We note in passing, that the identification of $N_\xi / G_\xi$ with $\mathcal{O}_\xi$ via

$$N_\xi / G_\xi \overset{\sim}{\leftarrow} G/G_\xi \overset{\sim}{\rightarrow} \mathcal{O}_\xi$$

is equal to the mapping from $N_\xi / G_\xi$ to $\mathcal{O}_\xi$, which is induced by the mapping $-\mu_{\tilde{R}'}$ from $\mathrm{T}^* G$ to $\mathfrak{g}^*$.

The mapping $G/G_\xi \to \mathcal{O}_\xi$ is given by a smooth injective immersion from $G/G_\xi$ to $\mathfrak{g}^*$, and the same is true for $-\mu_{\tilde{R}'} : N_\xi \to \mathfrak{g}^*$. In general one needs additional assumptions to ensure that this is an embedding. This is clearly the case if $G$ is compact; then the coadjoint orbits are smooth compact submanifolds of $\mathfrak{g}^*$.

If $G$ is compact, then one can identify $\mathfrak{g}$ with $\mathfrak{g}^*$ by means of an $\mathrm{Ad}\,G$-invariant inner product in $\mathfrak{g}$, and we get that the homogeneous spaces $G/G_\xi$ are the *(generalized) complex flag manifolds*, cf. [24, Section 2]. In the general case, in which the immersion need not be an embedding or, even if it is, the image need not to be closed, we will regard the coadjoint orbit $\mathcal{O}_\xi$ as an injectively immersed submanifold of $\mathfrak{g}^*$. The symplectic form on the coadjoint orbits (which among other implies that all coadjoint orbits have even dimension) was found by Lie [54, Kap. 19] and rediscovered by Kirillov [47, Lemma 5.2][1].

The inverses $\sigma_\xi^{-1}$ of the symplectic forms on the coadjoint orbits in $\mathfrak{g}^*$ piece together to a smooth Poisson structure in $\mathfrak{g}^*$. In general a *Poisson structure* in a smooth manifold $P$ is a smooth section $\varpi$ of $\Lambda^2\,\mathrm{T}\,P$ such that the corresponding brackets $\{f,\, g\} \in \mathrm{C}^\infty(P)$, defined by

$$\{f,\, g\}(p) := \varpi_p\,(\mathrm{d}f_p,\, \mathrm{d}g_p)\,,\; p \in P$$

for any $f,\, g \in \mathrm{C}^\infty(P)$, satisfy the Jacobi identity, turning $\mathrm{C}^\infty(P)$ into a Lie algebra. Such structures were introduced by Lie [54, Kap. 11] under the name of "function groups", and rediscovered by Lichnerowicz [53], who coined their name. The Poisson structure in $\mathfrak{g}^*$, found by Lie [54, Kap. 19], now is the one for which the Poisson brackets $\{f,\, g\}$ of $f,\, g \in \mathrm{C}^\infty\,(\mathfrak{g}^*)$ are equal to the restriction to $\mathfrak{g}^* = \mathrm{T}_1^*\,G \subset \mathrm{T}^*\,G$ of the Poisson brackets in $\mathrm{T}^*\,G$ of the unique $\tilde{L}$-invariant extensions of $f$ and $g$ to $\mathrm{T}^*\,G$. The smoothness of the Poisson structure in $\mathfrak{g}^*$ can be proved by using that the $\tilde{L}$-invariant extension of $f \in \mathrm{C}^\infty\,(\mathfrak{g}^*)$ to $\mathrm{T}^*\,G$ is equal to $-\,(\mu_{\tilde{R}'})^*\,f$. This also yields the Poisson identity for the brackets in $\mathrm{C}^\infty\,(\mathfrak{g}^*)$ as a consequence of the Jacobi identity for the Poisson brackets in $\mathrm{C}^\infty\,(\mathrm{T}^*\,G)$.

---

[1] Kirillov proved that the form is nondegenerate, whereas Kostant also proved that it is symplectic, cf. [50, Th.5.3.1].

Guillemin and Sternberg [36] used the symplectic structure $\sigma_{\mathcal{O}}$ of the coadjoint orbit $\mathcal{O}$ in $\mathfrak{g}^*$ in order to define the reduced phase space $(M_{\mathcal{O}}, \sigma_{\mathcal{O}})$ of the Hamiltonian $G$-action in $M$ *at the orbit* $\mathcal{O}$ as the reduced phase space at the level zero for the action of $G$ in $M \times \mathcal{O}$, which is defined by

$$g \cdot (m, \xi) = (g \cdot m, \operatorname{Ad}^* g(\xi)), \; g \in G, \; m \in M, \; \xi \in \mathcal{O}.$$

This action is Hamiltonian with momentum mapping given by

$$\mu_{M \times \mathcal{O}}(m, \xi) = \mu_M(m) - \xi, \; m \in M, \; \xi \in \mathcal{O}, \tag{15.19}$$

where the symplectic form in $M \times \mathcal{O}$ is taken to be equal to

$$\sigma_{M \times \mathcal{O}} = \sigma_M \oplus (-\sigma_{\mathcal{O}}).$$

For each $\xi \in \mathcal{O}$ we have that $\mathcal{O} = \mathcal{O}_\xi$ and the mapping $m \mapsto (m, \xi)$ induces a symplectic diffeomorphism from $(M_\xi, \sigma_\xi)$ onto $M_{\mathcal{O}}$ provided with its symplectic form as a reduced phase space. So the Marsden-Weinstein reduction at the level $\xi$ is isomorphic to the Guillemin-Sternberg reduction at the orbit $\mathcal{O} = \mathcal{O}_\xi$.

If $M$ and $G$ are compact, then the momentum mapping $\mu : M \to \mathfrak{g}^*$ of a Hamiltonian $G$-action on $M$ has a number of remarkable properties.

If $G = T$ is a torus, then Atiyah [3] and Guillemin and Sternberg [35] proved that the the set $M^T$ of fixed points of the $T$-action in $M$ is nonvoid, that its image $\mu\left(M^T\right)$ is a finite subset of $\mathfrak{t}^*$ and that the image $\mu(M)$ in $\mathfrak{t}^*$ of $M$ under $\mu$ is equal to the convex hull of $\mu\left(M^T\right)$. In particular, $\mu(M)$ is a convex polytope.

If $C$ is a connected component of the set of regular values in $\mathfrak{t}^*$ of $\mu$, then Duistermaat and Heckman [23] proved that the cohomology class

$$[\sigma_\xi] - \langle [\Omega_\xi], \xi \rangle \in \mathrm{H}^2_{\text{de Rham}}(M_\xi) \tag{15.20}$$

is constant as $\xi$ varies in $C$. Here $\Omega$ is the $\mathfrak{t}$-valued closed two-form in $M_\xi$ such that $\pi_\xi^* \Omega = \mathrm{d}\, \theta$, where $\theta$ is a connection form for the principal $T$-bundle

$\mu^{-1}(\{\xi\}) \to M_\xi$. This statement makes sense because the reduced phase spaces $M_\xi$ are diffeotopic, so the de Rham cohomology groups $\mathrm{H}^2_{\text{de Rham}}(M)$ are canonically isomorphic, if $\xi$ varies in $C$. The point is that $[\Omega_\xi]$ is constant as $\xi$ moves in $C$, because as a topological class it is integral.

This means that $[\sigma_\xi]$ depends in an affine linear way on $\xi \in C$, which has the consequence that the push-forward of the canonical measure $|\frac{1}{n!}\sigma^n|$ in $M$ to $\mathfrak{t}^*$ by means of $\mu$ has a density which in each component $C$ is equal to a polynomial. In turn this implies that the oscillatory integral

$$\int_M \mathrm{e}^{i\langle X,\, \mu\rangle}\, |\tfrac{1}{n!}\sigma^n| \tag{15.21}$$

is equal to the leading term of its stationary phase approximation as $X$ goes to infinity in $\mathfrak{g}$. Because the latter is expressed as a sum (integral) over the set of critical points of $\langle X,\, \mu\rangle$, which are the zeros of $X_M$, this can be viewed as a fixed point formula for (15.21). The proof in the case that the fixed point set has connected components of positive dimension has been given in the Addendum of [23], using ideas of Bott [13]

Berline and Vergne [10] and Atiyah and Bott [6] obtained the formula for (15.21) as an application of a much more general fixed point formula for integrals of equivariantly closed forms, as defined in Section 16.1. Actually, also the constancy of (15.20) gets its natural explanation in this framework. Indeed, the condition (15.12) for Hamiltonian actions just means that

$$\alpha(X) = \sigma - \langle X,\, \mu\rangle,\ X \in \mathfrak{g}$$

defines an equivariantly closed form of total degree 2. The antisymmetry of differential forms implies that, for any vector fields $v$ and $w$,

$$\mathrm{i}(v)\circ \mathrm{i}(w) + \mathrm{i}(w)\circ \mathrm{i}(v) = 0.$$

This implies that the Lie derivative is given by

$$\mathcal{L}(v) = \mathrm{d}\circ \mathrm{i}(v) + \mathrm{i}(v)\circ \mathrm{d} = \mathrm{d}_{\mathfrak{g}}\circ \mathrm{i}(v) + \mathrm{i}(v)\circ \mathrm{d}_{\mathfrak{g}},$$

hence $\mathrm{L}(v)\alpha = \mathrm{d}_{\mathfrak{g}}\,\mathrm{i}(v)\alpha$, because $\mathrm{d}_{\mathfrak{g}}\,\alpha = 0$. In our situation with $G = T$, $\mathfrak{g} = \mathfrak{t}$, we see as a consequence that the equivariant cohomology class

$$\left[\iota_{\xi}^{*}\alpha\right] \in \mathrm{H}_{G}^{2}\left(\mu^{-1}(\{\xi\})\right)$$

does not depend on $\xi \in C$. In view of the isomorphism of Theorem 16.1 of the equivariant cohomology of $\mu^{-1}(\{\xi\})$ with the de Rham cohomology of its $T$-orbit space $M_{\xi}$, which maps $\left[\iota_{\xi}^{*}\alpha\right]$ to (15.20), the constancy of the latter follows.

For a review on the relation between equivariant cohomology and Hamiltonian group actions, see [21].

If $G = K$ is a nonabelian compact group with maximal torus $T$, then Guillemin and Sternberg [35] conjectured that the intersection of $\mu_{K}(M)$ with each Weyl chamber in $\mathfrak{t}^{*}$ is also equal to a convex polytope, which they proved in some important special cases. Here $\mathfrak{t}^{*}$ is defined as the fixed point set in $\mathfrak{k}^{*}$ of the coadjoint action of $T$ in $\mathfrak{k}^{*}$. The proof in the general case has been given by Kirwan [48]. For further results, cf. Sjamaar [70].

## 15.3 The Complex Line Bundle

A background reference for this section and the next one is Kostant [50], where the constructions are part of the program of *geometric quantization* .

In this section we assume that $\sigma$ is a smooth two-form in $M$ such that its de Rham cohomology class is *integral*, in the sense that it is contained in the image of the Čech (sheaf) cohomology group $\mathrm{H}^{2}(M,\,\mathbf{Z})$ with coefficients in $\mathbf{Z}$, under the natural mappings

$$\mathrm{H}^{2}(M,\,\mathbf{Z}) \to \mathrm{H}^{2}(M,\,\mathbf{R}) \xrightarrow{\sim} \mathrm{H}^{2}_{\text{de Rham}}(M). \qquad (15.22)$$

Explicitly, this means the following. Let $\{U_{\alpha}\}_{\alpha\in A}$ be a covering of $M$ with open subsets such that the intersection of any finite subcollection, when nonvoid, is contractible. Such a covering can for instance be obtained by taking geodesic balls in normal coordinate neighborhoods, with respect to some Riemannian structure in $M$. The intersections of such balls are contractible because they are geodesically convex.

Since $\mathrm{d}\sigma = 0$, there exists in each $U_\alpha$ a smooth one-form $\tau_\alpha$ such that

$$\sigma = \mathrm{d}\tau_\alpha \ \text{ in } \ U_\alpha. \tag{15.23}$$

In $U_\alpha \cap U_\beta$ we have $\mathrm{d}\,(\tau_\alpha - \tau_\beta) = 0$, so there exists a smooth function $\mu_{\alpha\beta}$ in $U_\alpha \cap U_\beta$, determined modulo the addition of a constant, such that

$$\tau_\alpha - \tau_\beta = \mathrm{d}\mu_{\alpha\beta} \ \text{ in } \ U_\alpha \cap U_\beta. \tag{15.24}$$

here we have twice used Poincaré's lemma. By passing to $\frac{1}{2}\,(\mu_{\alpha\beta} - \mu_{\beta\alpha})$ if necessary, we can arrange that $\mu_{\beta\alpha} = -\mu_{\alpha\beta}$. In $U_\alpha \cap U_\beta \cap U_\gamma$, when nonvoid, we have the real-valued constant

$$\nu_{\alpha\beta\gamma} := \mu_{\alpha\beta} + \mu_{\beta\gamma} + \mu_{\gamma\alpha} \ \text{ in } \ U_\alpha \cap U_\beta \cap U_\gamma, \tag{15.25}$$

because

$$\mathrm{d}\nu_{\alpha\beta\gamma} = (\tau_\alpha - \tau_\beta) + (\tau_\beta - \tau_\gamma) + (\tau_\gamma - \tau_\alpha) = 0.$$

The $\nu_{\alpha\beta\gamma}$ form a two-cochain with values in the sheaf of real valued constants, the Čech cohomology class of which is mapped to the de Rham cohomology class of $\sigma$ by the second arrow in (15.22).

That $[\nu]$ belongs to the image of the first arrow in (15.22) now means that there exists a two-cochain of integers $c_{\alpha\beta\gamma}$, defined for all $\alpha,\ \beta,\ \gamma \in A$ such that $U_\alpha \cap U_\beta \cap U_\gamma \neq \emptyset$, and a one-cochain $d_{\alpha\beta} \in \mathbf{R}$, defined for all $\alpha,\ \beta \in A$ such that $U_\alpha \cap U_\beta \neq \emptyset$, for which

$$\mu_{\alpha\beta} + \mu_{\beta\gamma} + \mu_{\gamma\alpha} = \nu_{\alpha\beta\gamma} = c_{\alpha\beta\gamma} + d_{\alpha\beta} + d_{\beta\gamma} + d_{\gamma\alpha}. \tag{15.26}$$

That is, we could have replaced the $\mu_{\alpha\beta}$ by the $\mu_{\alpha\beta} - d_{\alpha\beta}$ in (15.24) and thereby arranged that all $\nu_{\alpha\beta\gamma}$ are integers. This implies that the functions

$$\phi_{\alpha\beta} := \mathrm{e}^{2\pi i \mu_{\alpha\beta}} \ \text{ in } \ U_\alpha \cap U_\beta \tag{15.27}$$

define a two-cochain with values in the sheaf of multiplicative groups of U(1)-valued functions, because

$$\phi_{\alpha\beta}\,\phi_{\beta\gamma}\,\phi_{\gamma\alpha} = 1 \ \text{ in } \ U_\alpha \cap U_\beta \cap U_\gamma. \tag{15.28}$$

Here U(1) is the multiplicative group of the $z \in \mathbf{C}$ such that $|z| = 1$.

The compatibility condition (15.28) implies that we can glue the $U_\alpha \times \mathbf{C}$ together along the intersections $U_\alpha \cap U_\beta$, by identifying $(m,\, z_\alpha) \in U_\alpha \times \mathbf{C}$ with $(m,\, z_\alpha) \in U_\beta \times \mathbf{C}$ if $m \in U_\alpha \cap U_\beta$ and $z_\alpha = \phi_{\alpha\beta}(m)\, z_\beta$. The result is a smooth complex line bundle $L$ over $M$, which inherits the Hermitian structure of $\mathbf{C}$, because the $\phi_{\alpha\beta}$ are U(1)-valued.

In this case the fiber dimension is equal to one, so the unitary frame bundle, which is a principal U(1)-bundle over $M$, is just equal to the bundle $\mathrm{U}(L)$ of the unit circles in the fibers. For every $z \in \mathrm{U}(1)$ we write $z_L$ for the action of multiplication by $z$ in the fibers of $L$. Similarly the infinitesimal action of $X \in \mathrm{u}(1) = i\mathbf{R}$ in $L$ is denoted by $X_L$. We will use this notation in particular for $X = 2\pi i$, the standard generator of the Lie algebra u(1) of U(1) such that $\mathrm{e}^X = 1$.

As defined in front of (6.7), a connection form in $\mathrm{U}(L)$ is a U(1)-invariant $i\mathbf{R}$-valued one-form $\theta$ in $\mathrm{U}(L)$, such that

$$\mathrm{i}\,(2\pi i_L)\,\theta = 2\pi i \quad \text{(constantly)}. \tag{15.29}$$

Let $\pi$ denote the projection from $\mathrm{U}(L)$ onto $M$ and let

$$z_\alpha : \pi^{-1}\,(U_\alpha) \to \mathrm{U}(1) \subset \mathbf{C}$$

denote the trivialization $\pi^{-1}\,(U_\alpha) \to U_\alpha \times \mathrm{U}(1)$, followed by the projection to the second factor. Note that the gluing condition means that

$$z_\alpha = \pi^*\phi_{\alpha\beta}\, z_\beta \ \text{ in } \ \pi^{-1}\,(U_\alpha \cap U_\beta)\,. \tag{15.30}$$

In $\pi^{-1}\,(U_\alpha)$ we get that

$$\theta = \pi^*\theta_\alpha + \tfrac{\mathrm{d}z_\alpha}{z_\alpha} \tag{15.31}$$

for a uniquely defined one-form $\theta_\alpha$. On the other hand we get in view of (15.30) that

$$\frac{\mathrm{d}z_\alpha}{z_\alpha} = \pi^*\left(\frac{\mathrm{d}\phi_{\alpha\beta}}{\phi_{\alpha\beta}}\right) + \frac{\mathrm{d}z_\beta}{z_\beta},$$

so we can use (15.31) as the definition of a connection form $\theta$ in $M$, if and only if

$$\pi^*\left(\theta_\alpha + \frac{\mathrm{d}\phi_{\alpha\beta}}{\phi_{\alpha\beta}}\right) = \pi^*\theta_\beta \text{ in } \pi^{-1}\left(U_\alpha \cap U_\beta\right),$$

which is equivalent to

$$\theta_\alpha + \frac{\mathrm{d}\phi_{\alpha\beta}}{\phi_{\alpha\beta}} = \theta_\beta \text{ in } U_\alpha \cap U_\beta, \tag{15.32}$$

because $\pi$ is a submersion.

Now (15.27) and (15.24) show that

$$\frac{\mathrm{d}\phi_{\alpha\beta}}{\phi_{\alpha\beta}} = 2\pi i\, \mu_{\alpha\beta} = 2\pi i\left(\tau_\alpha - \tau_\beta\right),$$

so we get that (15.32) is satisfied if we *define*

$$\theta_\alpha = -2\pi i\, \tau_\alpha \text{ in } U_\alpha. \tag{15.33}$$

For the thus defined connection form $\theta$ in UF(1) we get from (15.31) and the observation that $\frac{\mathrm{d}z_\alpha}{z_\alpha}$ is closed, that

$$-\frac{1}{2\pi i}\Omega = -\frac{1}{2\pi i}\,\mathrm{d}\theta = \mathrm{d}(\pi^*\tau_\alpha) = \pi^*\,\mathrm{d}\tau_\alpha = \pi^*\sigma,$$

where $\Omega$ is defined as in (6.7). Note that the Lie brackets in $\mathrm{u}(1)$ are zero, corresponding to the fact that $\mathrm{U}(1)$ is abelian. In other words, $\sigma$ is equal to the *Chern form* corresponding to the chosen connection form $\theta$ in $\mathrm{U}(L)$, cf. (13.6).

Reading the above identities backward, one gets that the Chern class of any connection in $\mathrm{U}(L)$, where $L$ is a Hermitian complex line bundle over $M$, is integral. So the final result is:

**Proposition 15.1** *The closed two-form $\sigma$ in $M$ is equal to the Chern form of a Hermitian connection $\theta$ in a complex line bundle $L$ over $M$, if and only if the de Rham cohomology class of $\sigma$ is integral.*

## 15.4 Lifting the Action

In this section we assume that $\sigma$ is a symplectic form in $M$ such that $[\sigma] \in \mathrm{H}^2_{\text{de Rham}}(M)$ is integral. Furthermore $G$ is a connected Lie group which acts on $M$ in a Hamiltonian fashion, with momentum mapping $\mu : M \to \mathfrak{g}^*$, which is determined modulo adding a constant $\nu$ in the fixed point space for the coadjoint action in $\mathfrak{g}^*$.

According to Proposition 15.1, the integrality condition is equivalent to the existence of a smooth complex line bundle $\pi : L \to M$, with a Hermitian structure and a connection form $\theta$ in the corresponding $\mathrm{U}(1)$-bundle $\mathrm{U}(L)$, such that

$$-\tfrac{1}{2\pi i}\, \mathrm{d}\theta = \pi^*\sigma. \tag{15.34}$$

We will now regard the Hermitian structure and the connection form $\theta$ as *part of the structure of the line bundle* $L$. As a consequence, an *automorphism of* $L$ will be a diffeomorphism $\Phi : L \to L$ which has the following properties i) — iii):

i) There exists a diffeomorphism $\phi$ of $M$ such that, for each $m \in M$, the restriction of $\Phi$ to $L_m$ is a complex linear mapping from $L_m$ to $L_{\phi(m)}$.

ii) $\Phi$ preserves the Hermitian structure in $L$, which means that $\Phi$ maps $\mathrm{U}(L)$ to itself. The restriction of $\Phi$ to $\mathrm{U}(L)$ will also be denoted by $\Phi$.

iii) $\Phi^*\theta = \theta$.

Note that the automorphism $\Phi$ is equivalently given by a diffeomorphism $\Phi$ of $\mathrm{U}(L)$ which commutes with the $\mathrm{U}(1)$-action and leaves $\theta$ invariant. The automorphisms of $L$ form a group, which will be denoted by $\mathrm{Aut}(L)$.

In the same fashion an *infinitesimal automorphism of* $L$ is defined as a $\mathrm{U}(1)$-invariant vector field $v$ in $\mathrm{U}(L)$ such that $\mathcal{L}(v)\theta = 0$. The infinitesimal automorphisms of $L$ form a Lie algebra of vector fields, which will be denoted by $\mathrm{End}(L)$.

Now a *lifting to* Aut$(L)$ *of the action of* $G$ *in* $M$ will be defined as a smooth homomorphism $g \mapsto g_L$ from $G$ to Aut$(L)$ such that, for each $g \in G$, $\pi \circ g_L = g_M \circ \pi$.

In the same fashion a *lifting to* End$(L)$ *of the infinitesimal action of* $\mathfrak{g}$ *in* $M$ will be defined as a homomorphism of Lie algebras $X \mapsto X_L$ from $\mathfrak{g}$ to End$(L)$ such that, for each $X \in \mathfrak{g}$, the projection $\pi$ intertwines $X_L$ with $X_M$.

Here the Lie algebra structure in End$(L)$ is the group-theoretic one, which is opposite to the Lie brackets of vector fields, cf. the comments in front of Lemma 9.1. So the homomorphism property means that

$$[X, Y]_L = -[X_L, Y_L], \tag{15.35}$$

where in the right hand side we have written the standard Lie brackets of vector fields.

Finally, if $v$ and $w$ are vector fields in manifolds $L$ and $M$, respectively, and $\pi : L \to M$ is a smooth map, then we say that $\pi$ *intertwines* $v$ *with* $w$ if

$$w_{\pi(l)} = \mathrm{T}_l\, \pi\, (v_l)\,, \; l \in L. \tag{15.36}$$

This is equivalent to the condition that, for every solution curve $\gamma$ in $L$ of $\gamma'(t) = v(\gamma(t))$, $\delta = \pi \circ \gamma$ is a solution curve in $M$ of the equation $\delta'(t) = w(\delta(t))$. This can be used to prove that, if $\pi$ intertwines $v_j$ with $w_j$, $j = 1, 2$, then $\pi$ intertwines $[v_1, v_2]$ with $[w_1, w_2]$.

Note that, for each vector field $w$ in $M$ there is a unique vector field $v$ in $\mathrm{U}(L)$ such that $\pi : \mathrm{U}(L) \to M$ intertwines $v$ with $w$ and $v$ is *horizontal* in the sense that $\mathrm{i}(v)\theta = 0$. We write $v = w^{\mathrm{hor}}$ in this case.

**Proposition 15.2** *A mapping* $X \mapsto X_L$ *from* $\mathfrak{g}$ *to the set of vector fields in* $\mathrm{U}(L)$ *is a lifting to* End$(L)$ *of the infinitesimal action* $X \mapsto X_M$ *of* $\mathfrak{g}$ *in* $M$ *if and only if there exists a momentum mapping* $\mu : M \to \mathfrak{g}^*$ *for the Hamiltonian action in* $M$*, such that*

$$X_L = X_M^{\mathrm{hor}} - \pi^* \langle X, \mu \rangle\, 2\pi i_L, \quad X \in \mathfrak{g}. \tag{15.37}$$

**Proof** The condition that $\pi$ intertwines $X_L$ with $X_M$ means that there exists a smooth real-valued function $f_X$ on $L$ such that $X_L = X_M^{\text{hor}} + f_X\, 2\pi i_L$, and in this case $X_L$ is U(1)-invariant if and only if $f_X$ is constant on the fibers, which means that $f_X = \pi^* \phi_X$ for some smooth function $\phi_X$ on $M$. The condition that $X \mapsto X_L$ is a linear mapping now is equivalent the existence of a smooth mapping $\phi : M \to \mathfrak{g}^*$ such that, for every $X \in \mathfrak{g}$, $\phi_X = \langle X, \phi \rangle$. So from now on we may restrict our attention to

$$X_L = X_M^{\text{hor}} + \pi^* \langle X, \phi \rangle\, 2\pi i_L, \quad X \in \mathfrak{g}. \tag{15.38}$$

The invariance of $\theta$ under $X_L$ is equivalent to

$$\begin{aligned} 0 &= \mathcal{L}(X_L)\,\theta = \mathrm{i}\,(X_L)\,\mathrm{d}\theta + \mathrm{d}\,(\mathrm{i}\,(X_L)\,\theta) \\ &= -\,\mathrm{i}\,(X_L)\; 2\pi i\, \pi^* \sigma + 2\pi i\; \mathrm{d}\,\pi^* \langle X, \phi \rangle \\ &= 2\pi i\, \pi^* \left( -\,\mathrm{i}\,(X_M)\,\sigma + \mathrm{d}\langle X, \phi \rangle \right) \\ &= 2\pi i\, \pi^* \left( \mathrm{d}\langle X, \mu \rangle + \mathrm{d}\langle X, \phi \rangle \right), \end{aligned}$$

where we have used (15.34) and (15.29) to get to the second line. This means that $\mathrm{d}\langle X, \mu \rangle + \mathrm{d}\langle X, \phi \rangle = 0$, because $\pi$ is a submersion, or $-\phi = \mu + \nu$ for a *constant* element $\nu$ of $\mathfrak{g}^*$.

In order to compute the Lie brackets of the vector fields $X_L$ and $Y_L$ for $X, Y \in \mathfrak{g}$, we write

$$X_L^{\text{hor}} = X_M^{\text{hor}} \text{ and } X_L^{\text{vert}} = \pi^* \langle X, \phi \rangle\, 2\pi i_L$$

for the horizontal and vertical part of $X_L$, respectively.

In order to compute $\left[ X_L^{\text{hor}}, Y_L^{\text{hor}} \right]$, we start by observing that $\pi$ intertwines it with $[X_M, Y_M] = -[X, Y]_M$, so there is a smooth real-valued function $f$ on U($L$), such that

$$\left[ X_L^{\text{hor}}, Y_L^{\text{hor}} \right] = -[X, Y]_M^{\text{hor}} + f\, 2\pi i_L.$$

Applying $\theta$ to both sides and using (2.16), we get

$$\begin{aligned} f &= -\tfrac{1}{2\pi i}\, \mathrm{d}\theta\left(X_L^{\mathrm{hor}},\, Y_L^{\mathrm{hor}}\right) = \pi^*\sigma\left(X_L^{\mathrm{hor}},\, Y_L^{\mathrm{hor}}\right) \\ &= \pi^*\left(\sigma\left(X_M,\, Y_M\right)\right) = -\pi^*\langle [X,\, Y],\, \mu\rangle, \end{aligned}$$

where in the last identity we used (15.15). The conclusion therefore is that

$$\left[X_L^{\mathrm{hor}},\, Y_L^{\mathrm{hor}}\right] = -[X,\, Y]_M^{\mathrm{hor}} - \pi^*\langle [X,\, Y],\, \mu\rangle\, 2\pi i_L.$$

In a similar fashion we get that $\pi$ intertwines $\left[X_L^{\mathrm{hor}},\, Y_L^{\mathrm{hor}}\right]$ with $[X_M,\, 0] = 0$, so there is a smooth real-valued function $f$ on $\mathrm{U}(L)$ such that

$$\left[X_L^{\mathrm{hor}},\, Y_L^{\mathrm{vert}}\right] = f\, 2\pi i_L.$$

Applying $\theta$ and using (2.16), we get this time

$$\begin{aligned} f &= \pi^*\sigma\left(X_L^{\mathrm{hor}},\, Y_L^{\mathrm{vert}}\right) + X_L^{\mathrm{hor}}\pi^*\langle Y,\, \phi\rangle = \pi^*\left(X_M\langle X,\, \phi\rangle\right) \\ &= -\pi^*\left(X_M\langle X,\, \mu\rangle\right) = \pi^*\langle [X,\, Y],\, \mu\rangle, \end{aligned}$$

where in the last identity we used that $\mu$ intertwines the action of $G$ in $M$ with the coadjoint action in $\mathfrak{g}^*$.

Finally $X_L^{\mathrm{vert}}$ and $Y_L^{\mathrm{vert}}$ are tangent to the fibers and on each fiber equal to a constant multiple of $2\pi i_L$, so they commute. The conclusion is that

$$\begin{aligned} [X_L,\, Y_L] &= \left[X_L^{\mathrm{hor}},\, Y_L^{\mathrm{hor}}\right] + \left[X_L^{\mathrm{hor}},\, Y_L^{\mathrm{vert}}\right] - \left[Y_L^{\mathrm{hor}},\, X_L^{\mathrm{vert}}\right] \\ &= -[X,\, Y]_M^{\mathrm{hor}} + \pi^*\langle -[X,\, Y] + [X,\, Y] - [Y,\, X],\, \mu\rangle\, 2\pi i_L \\ &= -[X,\, Y]_M^{\mathrm{hor}} + \pi^*\langle [X,\, Y],\, \mu\rangle\, 2\pi i_L \end{aligned}$$

We recognize that that is equal to

$$-[X,\, Y]_L = -[X,\, Y]_M^{\mathrm{hor}} - \pi^*\langle [X,\, Y],\, \phi\rangle\, 2\pi i_L$$

if and only if $\langle [X,\, Y],\, \nu\rangle = 0$. So we get that $X \mapsto X_L$ is a lift to $\mathrm{End}(L)$, if and only if $\nu \in [\mathfrak{g},\, \mathfrak{g}]^0$, which means that $\mu + \nu$ is a momentum map for the action of $G$ in $M$. □

For the lifting to $\mathrm{Aut}(L)$ of the $G$-action in $M$, we recall that the universal covering group of $G$ is a connected Lie group $\tilde{G}$, together with a surjective homomorphism $\tilde{\pi} : \tilde{G} \to G$, such that $\Gamma := \ker \tilde{\pi}$ is a discrete subgroup of $\tilde{G}$ which is canonically isomorphic to the fundamental group $\pi_1(G, 1_G)$ of $G$ with base point equal to the identity element of $G$. This implies that $\mathrm{T}_1 \tilde{\pi}$ is an isomorphism from the Lie algebra $\tilde{\mathfrak{g}}$ of $\tilde{G}$ onto $\mathfrak{g}$, this will be used to identify $\tilde{\mathfrak{g}}$ with $\mathfrak{g}$. It is also known that $\Gamma$ is contained in the center of $\tilde{G}$, which means that each element of $\Gamma$ commutes with each element of $\tilde{G}$; in particular $\Gamma$ is abelian. Each subgroup $\Delta$ of $\Gamma$ leads to an intermediate covering

$$\tilde{G} \to \tilde{G}/\Delta \to G$$

of $G$.

**Proposition 15.3** *Let $U$ be a torus and $P$ a principal $U$-bundle over a connected manifold $M$, with connection form $\theta$. Let $G$ be a connected Lie group which acts on $M$ and suppose that $X \mapsto X_P$ is a homomorphism from $\mathfrak{g}$ to* $\mathrm{End}(P, \theta)$*, which is intertwined by the projection $\pi : P \to M$ with the infinitesimal action $X \mapsto X_M$ of $\mathfrak{g}$ in $M$. Then there exists a unique homomorphism $\tilde{g} \mapsto \tilde{g}_P$ from $\tilde{G}$ to* $\mathrm{Aut}(P, \theta)$*, the infinitesimal action of which is equal to $X \mapsto X_P$.*

*Furthermore, $\Gamma$ acts via $U$ in the sense that there exists a homomorphism $\mathcal{M} : \Gamma \to U$ such that, for each $\gamma \in \Gamma$, we have that $\gamma_P = \mathcal{M}(\gamma)_P$. The action of $G$ in $M$ can be lifted to an action of the covering $\tilde{G}/\ker \mathcal{M}$ of $G$ by means of automorphisms of $(P, \theta)$.*

**Proof** The elements $\tilde{g}$ in $\tilde{G}$ which cover $g \in G$ are defined as the group of homotopy classes of smooth curves $\gamma : [0, 1] \to G$ such that $\gamma(0) = 1_G$ and $\gamma(1) = g$.

If the action of $G$ in $M$ would lift to $\mathrm{Aut}(P)$, with infinitesimal action equal to $X \mapsto X_P$, then we would have, for any $p \in P$, that $p(t) = \gamma(t)_P \cdot p$

satisfies the differential equation with initial condition

$$\tfrac{dp(t)}{dt} = X(t)_{P,\, p(t)}, \quad p(0) = p, \tag{15.39}$$

where $X(t) \in \mathfrak{g}$ is defined by

$$X(t) = \tfrac{d}{ds}\gamma(s)\,\gamma(t)^{-1}|_{s=t}.$$

The existence and uniqueness theorem for ordinary differential equations yields that (15.39) has a unique maximal solution. If it is not defined for all $[0,\, 1]$, so its domain of definition is of the form $[0,\, t^*[$ for some $t^* \in ]0,\, 1]$, then $p(t)$ runs to infinity in the sense that for each compact subset $K$ of $P$ there exists $0 \leq t < t^*$ such that $p(t) \notin K$. However, both curves $\pi(p(t))$ and $\gamma(t)_M \cdot \pi(p)$ in $M$ satisfy the differential equation with initial condition

$$\tfrac{dm(t)}{dt} = X(t)_{M,\, m(t)}, \quad m(0) = \pi(p), \tag{15.40}$$

because $\pi$ interwines the infinitesimal actions. Therefore we get that $\pi(p(t)) = \gamma(t)_M \cdot \pi(p)$ for all $0 \leq t < t^*$. Because the projection $\pi : P \to M$ is a proper mapping, this leads to a contradiction with the running to infinity of $p(t)$. The conclusion is that (15.39) has a unique maximal solution $p : [0,\, 1] \to P$; we write $p(1) = \gamma_P \cdot p$.

The homomorphism property of $X \mapsto X_P$ yields that $p(1)$ is constant under an homotopy of $\gamma$, preserving initial and end points. In this way, we get the desired lifting of the action of $G$ in $M$ to an action $\tilde{g} \mapsto \tilde{g}_P$ of $\tilde{G}$ on $P$, by means of automorphisms of $(P,\, \theta)$. Now let $\tilde{g} \in \ker \tilde{\pi}$. Then $\pi \circ \tilde{g}_P = \tilde{\pi}(\tilde{g})_M \circ \pi = \pi$, so $\tilde{g}_P$ leaves the fibers of $P$ invariant. Because it commutes with the action of $U$, we conclude that for each $m \in M$ there is a unique element $u(m) \in U$, such that $\tilde{g}_P \cdot p = u(m)_P \cdot p$ if $\pi(p) = m$. The statement about the homomorphism $\mathcal{M}$ however means that $u(m)$ does not depend on $m \in M$, which we will prove now.

For this purpose, we introduce a Riemannian structure $\beta$ in $M$. In each horizontal space $\ker \theta_p$ we get the inner product which is equal to the pull-back of $\beta_{\pi(p)}$ by means of the linear isomorphism $\mathrm{T}_p\,\pi : \ker \theta_p \to \mathrm{T}_{\pi(p)}\, M$.

Let $\beta_{\mathfrak{u}}$ be an inner product in the Lie algebra $\mathfrak{u}$ of $U$. We transplant this inner product to the tangent space of the fibers (which are the $U$-orbits), by means of the infinitesimal action of $\mathfrak{u}$ in $P$. We then get a unique inner product $\beta_p$ in $\mathrm{T}_p P$ for which the tangent space to the fiber and the horizontal space are orthogonal and which on each of these spaces is equal to the previously defined inner products, as in (9.3).

The point is that $\tilde{g}^* \theta = \theta$ implies that $\tilde{g}$ leaves the Riemannian structure $\beta$ in $P$ invariant. It is known that an isometry $I$ for a Riemannian structure is entirely determined, for any point $p$, by $I(p)$ and $\mathrm{T}_p I$. For instance by using that $I$ maps geodesics to geodesics, we get that $I$ is determined in a normal coordinate neighborhood, and then the global uniqueness follows from the local uniqueness beause $P$ is connected. Choose $p \in P$, let $u \in U$ be the element such that $\tilde{g} \cdot p = u_P \cdot p$. Also write $m \in M$ and for each $v \in \mathrm{T}_m M$, $v_p^{\mathrm{hor}}$ for the unique horizontal vector which is projected to $v$. The fact that, for each $v \in \mathrm{T}_m M$,

$$\mathrm{T}_p \tilde{g} \left( v_p^{\mathrm{hor}} \right) = v_{u_P \cdot p}^{\mathrm{hor}} = \mathrm{T}_p u_P \left( v_p^{\mathrm{hor}} \right),$$

and that on the tangent space of the fiber $\mathrm{T}_p \tilde{g}$ and $\mathrm{T}_p u_P$ coincide because both $\tilde{g}_P$ and $u_P$ commute withe the $U$-action, now implies that $\tilde{g} = u_P$. □

We will apply Proposition 15.3 to the case that $P = \mathrm{U}(L), U = \mathrm{U}(1)$, and $X_P$ is equal to the $X_L$ in (15.37). The homomorphism $\mathcal{M} : \pi_1(G, 1_G) \to \mathrm{U}(1)$ could be called the *monodromy character* of the infinitesimal action $X \mapsto X_L$.

In Section 15.5 we will be interested in the special case that the group $G$ is compact. In that case we have:

**Proposition 15.4** *If $G = T$ is a torus, then there exists a lift to $\mathrm{Aut}(L)$ of the Hamiltonian action of $T$ in $M$. If $G = K$ is a compact Lie group, then there exists a finite covering $\hat{K}$ of $K$ such that the same conclusion holds with $T$ replaced by $\hat{K}$.*

**Proof** If $G = T$, then we choose a basis $X(j)$ of the lattice of $X \in \mathfrak{t}$ such that $e^X = 1$. Note that the closed curves $t \mapsto e^{tX(j)}$, $t \in [0, 1]$, generate the fundamental group $\pi_1(T, 1)$. Let $X_L$ be defined as in (15.37). For any $\nu \in \mathfrak{t}^*$ we may replace $\mu$ by $\mu + \nu$ in (15.37); we denote the resulting new vector field in $\mathrm{U}(L)$ by $\tilde{X}_L$. Then the flows are related by

$$e^{\tilde{X}_L} = e^{X_L} \, e^{-2\pi i \langle X, \nu \rangle} .$$

For each $j$, we have $e^{X(j)} = 1$, which in $\tilde{G}$ means that $e^{X(j)} \in \Gamma = \pi_1(G, 1)$. write

$$\lambda(j) = \mathcal{M}\left(e^{X(j)}\right) \in \mathrm{U}(1),$$

which just means that

$$e^{X(j)_L} = \lambda(j)_L .$$

We now choose $\nu(j) \in \mathbf{R}$ such that

$$e^{2\pi i \nu(j)} = \lambda(j);$$

note that $\nu(j)$ is determined up to the addition of an integer. Let $\nu \in \mathfrak{t}^*$ be the unique element such that, for each $j$,

$$\langle X(j), \nu \rangle = \nu(j).$$

We then get for each $j$ that $e^{\widetilde{X(j)}_L} = 1$. So, if we replace $\mu$ by the momentum mapping $\mu + \nu$ in (15.37), then we get that the monodromy character is trivial and the first conclusion follows.

For the second conclusion one uses the fact that for the semisimple Lie algebra $\mathfrak{k}' = [\mathfrak{k}, \mathfrak{k}]$ the corresponding Lie subgroup $K'$ of $K$ has a finite fundamental group, so $\widetilde{K'} \to K'$ is a finite covering. On the other hand, the center $Z$ of $K$ is a torus, with Lie algebra $\mathfrak{z}$, and $\mathfrak{k} = \mathfrak{k}' \oplus \mathfrak{z}$. By a suitable neutral shift we get a lift to $\mathrm{Aut}(L)$ of the action of $Z$ in $M$, we also have a lift to $\mathrm{Aut}(L)$ for the action of the simply connected $\widetilde{K'}$. Combining these two, we get a lift to $\mathrm{Aut}(L)$ of the group $\hat{K} = \widetilde{K'} \times Z$. The map

$$(k, z) \mapsto \tilde{\pi}(k) \cdot z$$

is a finite covering $\hat{K} \to K$; it is a homomorphism because each element of $Z$ commutes with each element of $K'$. □

## 15.5 The Spin-c Dirac Operator

In this section, $M$ will be a compact and connected manifold with a symplectic form $\sigma$, the de Rham cohomology class of which is integral, cf. Section 15.3. This condition just means that there exists a complex line bundle $L$ over $L$ with a Hermitian structure and a connection form $\theta$ on the corresponding unitary frame bundle, such that (15.34) holds. Furthermore we assume that $K$ is a compact and connected Lie group, which acts in a Hamiltonian fashion on $M$.

According to Proposition 15.4, after passing to a finite covering of $K$ if necessary, there is a lift $k \to k_L$ to $\mathrm{Aut}(L, \theta)$ of the action of $K$ in $M$. Let $\mu : M \to \mathfrak{k}^*$ be the momentum mapping for the Hamiltonian action of $K$ in $M$, such that the infinitesimal action of $X \in \mathfrak{k}$ in $\mathrm{U}(L)$ is given by (15.37).

For the definition of the spin-c Dirac operator acting on sections of $E \otimes L$, we also need a $K$-invariant smooth almost complex structure $J$ in $M$, such that (2.24) defines a Riemannian structure $\beta$ in $M$. Then $h = \beta + i\,\sigma$ is a $K$-invariant Hermitian structure in $M$. As usual $K^*$ denotes the dual of the canonical line bundle of the almost complex manifold $M$, where we apologize for the clash of notation with the compact group $K$. Because a $K$-invariant Hermitian connection in $K^*$ is readily obtained by averaging an arbitrary Hermitian connection in $K^*$ over $K$, we then have all the required structures for the definition of the spin-c Dirac operator.

For the construction[1] of a $K$-invariant smooth almost complex structure $J$ in $M$ one may start with a $K$-invariant Riemannian structure $\beta^0$ in $M$, obtained by averaging an arbitrary such one over $K$. Then consider in

[1] I am grateful to Alan Weinstein for reminding me of this construction.

each tangent space $\mathrm{T}_x M$ the linear transformation $A_x = {\sigma_x}^{-1} \circ \beta_x^0$. It is antisymmetric with respect to $\beta_x$, so $-{A_x}^2$ is symmetric and positive definite. There is a unique positive definite symmetric $P_x$, such that ${P_x}^2 = -{A_x}^2$; it depends smoothly on $x$. Explicitly, there is an orthonormal basis on which the matrix of $A_x$ consists of $2 \times 2$-matrices

$$\begin{pmatrix} 0 & -p_j \\ p_j & 0 \end{pmatrix}$$

along the diagonal, with $p_j > 0$. On this basis the matrix of $P_x$ consists of $2 \times 2$-matrices

$$\begin{pmatrix} p_j & 0 \\ 0 & p_j \end{pmatrix}.$$

It follows that $P_x$ and $A_x$ commute, so that $J_x := A_x \, {P_x}^{-1}$ satifies

$$ {J_x}^2 = {A_x}^2 \left({P_x}^2\right)^{-1} = -1. \tag{15.41}$$

Also note that

$$\beta_x := \sigma_x \circ J_x = \beta_x^0 \circ P_x$$

is positive definite, so it defines a Riemannian structure in $M$.

All the constructions can be extended to the case of symplectic orbifolds, which we will often need when $(M, \sigma)$ itself is equal to the reduced phase space for some Hamiltonian group action. For instance, we may have a symplectic manifold $(\check{M}, \check{\sigma})$ on which we have a Hamiltonian action of a connected Lie group $H$, with momentum mapping $\check{\mu} : \check{M} \to \mathfrak{h}^*$. If $N$ is a normal connected Lie subgroup of $H$, then it acts in a Hamiltonian fashion on $\check{M}$, with momentum mapping $\mu_N = \rho \circ \check{\mu}$, where $\rho : \mathfrak{h}^* \to \mathfrak{n}^*$ denotes the restriction of linear forms on $\mathfrak{h}$ to linear forms on $\mathfrak{n}$. Note that, since $N$ is a normal subgroup of $H$, the adjoint action of $H$ on $\mathfrak{h}$ leaves $\mathfrak{n}$ invariant, so it induces a linear action on $\mathfrak{n}$ which we denote by $\mathrm{Ad}_{\mathfrak{n}}$. The dual action of $H$ on $\mathfrak{n}^*$ will be denoted by $\mathrm{Ad}_{\mathfrak{n}}^*$ and, for each $\nu \in \mathfrak{n}^*$, we have the stabilizer subgroup

$$H_\nu := \{h \in H \mid \mathrm{Ad}_{\mathfrak{h}}^*(\nu) = \nu\}$$

of $\nu$ in $H$. Note that $H_\nu$ leaves ${\mu_N}^{-1}(\{\nu\})$ invariant and that $N_\nu$ is a normal subgroup of $H_\nu$. So, if $\nu$ is a regular value of $\mu_N$ and the action of $N_\xi$ on ${\mu_N}^{-1}(\{\nu\})$ is proper, then we get a reduced phase space

$$M := {\mu_N}^{-1}(\{\nu\})/N_\nu$$

which is an orbifold, and the action of $H_\nu$ on ${\mu_N}^{-1}(\{\nu\})$ induces an action of $K := H_\nu/N_\nu$ on $M$. Using the arguments around (15.17), we get that the action of $K$ on $M$ is Hamiltonian, with momentum mapping $\mu$, such that

$$\pi_\nu^*\mu = (\iota_\nu^*\check{\mu})_{\mathfrak{h}_\nu},$$

where the subscript $\mathfrak{h}_\nu$ means that at each point of ${\mu_N}^{-1}(\{\nu\})$ the linear form on $\mathfrak{h}$ is restricted to $\mathfrak{h}_\nu$. The latter can be viewed as a linear form on the Lie algebra $\mathfrak{h}_\nu/\mathfrak{n}_\nu$ of $H_\nu/N_\nu$, because it is constant on the translates of $\mathfrak{n}_\nu$ in $\mathfrak{h}_\nu$.

A seemingly trivial example is obtained by taking $\check{M} = \mathbf{C}^d$, which is a Kähler manifold when provided with the standard hermitian metric. In it we have the action of the $d$-dimensional torus $\mathrm{U}(1)^d$, where $t \in \mathrm{U}(1)^d$ acts by sending $z \in \mathbf{C}^d$ to the vector $t \cdot z$ such that, for each $1 \leq j \leq d$, $(t \cdot z)_j = t_j\, z_j$. This action is Hamiltonian, with momentum mapping

$$\check{\mu} : z \mapsto \tfrac{1}{2}\left(|z_1|^2, \ldots, |z_d|^2\right).$$

Now let $N$ be any subtorus of $\mathrm{U}(1)^d$. It is a nice exercise to investigate the conditions that $\nu \in \mathfrak{n}^*$ is a regular value of $\mu_N$ and that ${\mu_N}^{-1}(\{\nu\})$ is compact. We then get a reduced phase space which is a compact symplectic orbifold, on which the torus $\mathrm{U}(1)^d/N$ acts in a Hamiltonian fashion. These objects are known in algebraic geometry as *toric varieties*. For a discussion of these from the point of view of this section, see [34].

Of course, one wants to choose a level at which to perform the reduction which is such that the de Rham cohomology class of the symplectic form of the reduced phase space is integral. In the investigation of this question the constancy of (15.20) as $\xi$ varies in $C$ will be a useful tool.

In the application of Theorem 14.1 to our situation, we get the simplification that there is no summation over $j$ in (14.4), because the fibers of $L$ are one-dimensional. Moreover, because the Chern form of $L$ is equal to the symplectic form $\sigma$, we get

$$\mathrm{ch}(L) = \mathrm{e}^{\sigma}, \tag{15.42}$$

cf. (13.10). That is, this factor in the fixed point formula is expressed directly in terms of the symplectic structure of $M$. The Todd form on the other hand is a characteristic form for the almost complex structure $J$. Note that if for instance we multiply $\sigma$ by a positive integer $m$, then the Chern character changes whereas we can keep the same $J$, hence the same Todd form.

For the application of Theorem 14.1 with $\gamma_M = k_M, \gamma_L = k_L, k \in K$, we make one more preparation. Let $T$ be a maximal torus in $K$, with Lie algebra $\mathfrak{t}$, so that $T = \exp \mathfrak{t}$. It is known that every element of of $K$ is conjugate to an element of $T$. Because traces don't change under conjugations, it therefore sufficies to apply Theorem 14.1 with $\gamma$ replaced by $t = \mathrm{e}^X$, $X \in \mathfrak{t}$. For $t$ in the complement of finitely many translates of subtori, we get that the fixed point set of $t_M$ in $M$ is equal to the fixed point set $M^T$ of the $T$-action, and there are no resonances with the orbifold structure. Note that if $x \in M^T$, then $\mathrm{e}^{X_L}$ acts as multiplication by the factor $\mathrm{e}^{-2\pi i \langle X, \mu \rangle}$ in the fiber $L_x$. That is, the factors $\lambda_{\tilde{L}, j}$ in (14.4) get replaced by the factor

$$\lambda_{\tilde{L}} = \epsilon_{\tilde{L}} \, \mathrm{e}^{-2\pi i \langle X, \mu \rangle}, \tag{15.43}$$

where $\epsilon_{\tilde{L}}$ is a root of unity which is independent of $X \in \mathfrak{t}$ and which comes from the orbifold strucure, so we have $\epsilon_{\tilde{L}} = 1$ in the smooth case. See Remark 2 after Theorem 14.1. The interesting point here is that the factor from the action on $L$ is expressed in terms of the value of the momentum $\mu$ at the connected component $F$ of $M^T$. Note that $\mu$ is constant on $F$, because the zeros of $X_M$ are the critical points of the hamilton function $\langle X, \mu \rangle$ of $X_M$ so, $\mathrm{d}\langle X, \mu \rangle = 0$ along $F$.

The factors in the denominator in (14.4) involve the action of $T$ on the normal bundles of the fixed point orbifolds, which is related to the Hessian of the Hamiltonian functions. The zeros of these factors are responsible for singularities in the contribution of the term which comes from a given fixed point orbifold. That is, each such term is a rational function on $T$ with poles along finitely many translates of subtori. Of course, in the sum over all the fixed point orbifolds the poles cancel, because we are dealing with the difference of the characters of two finite-dimensional representations.

An interesting special case occurs if the almost complex structure $J$ can be chosen to be integrable, in which case we have a complex analytic manifold which is Kähler, because $\sigma$, the imaginary part of the Hermitian structure, is closed. (The Kähler case can be viewed as the intersection of the complex analytic case and the symplectic case. I don't know enough about the non-Kähler complex analytic case and the non-Kähler symplectic case to make any sensible comparison.) Also, the line bundle $L$ then can be chosen to be holomorphic. See [75, Ch. III, Sec. 4]; note that $\sigma$ is a differential form of type $(1, 1)$, cf. loc. cit., p. 180. In this situation, Kodaira's embedding theorem, loc. cit. Ch. VI, says that actually $M$ can be viewed as a projective complex algebraic variety, in the sense that there is a complex analytic embedding from $M$ into a complex projective space; by Chow's theorem the image then is a complex algebraic subvariety. The "gaga principle" implies that the holomorphic line bundle $L$ is algebraic as well. See Serre [69, Sec. 20]. Conversely, the projective complex algebraic varieties, with algebraic line bundles over these, form a very rich and intensively studied subject in algebraic geometry.

In the Kähler case Guillemin and Sternberg [36] considered the action of $K$ on the space $H = \mathrm{H}^0(M, \mathcal{O}(L))$ of holomorphic sections of $L$ over $M$. An important special case in question are the coadjoint orbits, the complex flag manifolds. The famous theorem of Borel-Weil [68] and Tits [71] says that the representations of $K$ in the spaces of holomorphic sections of the line bundles over the complex flag manifolds are in bijective correspondence withe the equivalence classes $\rho$ of irreducible representations of $K$, parametrized by Élie Cartan by means of the highest weights. Atiyah and Bott [6, Section 5] verified that the character formula of Hermann Weyl coincides with the holomorphic Lefschetz fixed point formula, noting that the higher cohomology groups with values in the sheaf $\mathcal{O}(L)$ vanish in this case.

Guillemin and Sternberg [36] showed that if zero is a regular value of the momentum mapping (and the action of $K$ on the zero level set is free), then the reduced phase space $M_0$ is Kähler too, with a holomorphic line bundle $L_0$. Moreover, they found a natural isomorphism of the space of $K$-invariant holomorphic sections of $L$ over $M$ with the space of *all* holomorphic sections of $L_0$ over $M_0$. Performing the reduction at a coadjoint orbit which satisfies the integrality condition and which gives rise to the irreducible representation $\rho$, one get a natural identification of the subspace of $H$ of type $\rho$ with the space of all holomorphic sections of the line bundle over the reduced space.

As Atiyah and Bott [6, Sec. 5] did for the complex flag manifolds, one can use the holomorphic Lefschetz fixed point formula in $M$ to compute the character of the representation of $K$ in $H$, if the higher cohomology groups with values in the sheaf $\mathcal{O}(L)$ vanish. The latter is the case for instance if the line bundle $K^* \otimes L$ is positive, this is Kodaira's vanishing theorem, cf. the Remark after Proposition 6.1. The dimension of the space of $K$-invariant sections is equal to the average of the character over $K$, so it can in principle be determined from the fixed point formula.

This opens the way for generalizing the results of Guillemin and Sternberg [36] to arbitrary symplectic orbifolds, which need not be complex analytic (Kähler). In fact, it has been proved by Meinrenken [58] in full generality, after results by various authors in important special cases, that the average over $K$ of the virtual character always is equal to the difference of the dimensions of $\ker D_0^+$ and $\ker_0^-$, where $D_0^{\pm}$ is the spin-c Dirac operator of the reduced phase space $M_0$ at the level zero. Note that for this difference we have an integral formula in terms of the symplectic form $\sigma_0$ in $M_0$.

If $\mathcal{O}$ is an integral coadjoint orbit, with corresponding irreducible representation $\rho$, then one can apply this to $M \times \mathcal{O}$, with the momentum mapping as in (15.19). The conclusion then is that the difference of the multiplicities of $\rho$ in $\ker D^{+}$ and in $\ker D^{-}$ is equal to the difference of the dimensions of $\ker D^{+}_{\mathcal{O}}$ and $\ker^{-}_{\mathcal{O}}$, where $D^{\pm}_{\mathcal{O}}$ is the spin-c Dirac operator of the reduced phase space $M_{\mathcal{O}}$ at the orbit $\mathcal{O}$.

# Chapter 16

# Appendix: Equivariant Forms

## 16.1 Equivariant Cohomology

Equivariant cohomology is a structure which is attached to a smooth effective action of a Lie group $G$ on a smooth manifold $P$. The model of Henri Cartan [16] for it is a variation of de Rham cohomology, in which the algebra $\Omega^*(P)$ of smooth complex-valued differential forms on $P$ is replaced by the algebra

$$A := (\mathbf{C}[\mathfrak{g}] \otimes \Omega^*(P))^G \tag{16.1}$$

of *$G$-equivariant polynomial mappings* $\alpha$

$$\mathfrak{g} \ni X \mapsto \alpha(X) \in \Omega^*(P) \tag{16.2}$$

from the Lie algebra $\mathfrak{g}$ of $G$ to $\Omega^*(P)$. The equivariance of $\alpha$ means that

$$\alpha\left(\operatorname{Ad} g(X)\right) = \left(g_P^*\right)^{-1}\left(\alpha(X)\right), \quad g \in G, \ X \in \mathfrak{g}. \tag{16.3}$$

Here, for $g \in G$, $\operatorname{Ad} g$ is the adjoint action of conjugation by $g$ on the Lie algebra $\mathfrak{g}$, and $g_P^*$ denotes the pullback of differential forms by means of the action $g_P : P \to P$ of $g$ on $P$.

If $\alpha$ is a constant, as a differential form on $P$, then $\alpha$ is a conjugacy invariant polynomial on $\mathfrak{g}$. On the other hand, if $\alpha$ is a constant as a polynomial on $\mathfrak{g}$, then $\alpha$ is a $G$-invariant differential form on $P$. In this way, the algebra $A$ in (16.1) is a natural framework for the objects which appear in the Weil homomorphism, cf. Theorem 13.1.

In $A$ an operator $\mathrm{d}_{\mathfrak{g}}$ is introduced such that the range is contained in the kernel; the quotient of the kernel and the range is called the equivariant cohomology $\mathrm{H}_G^*(P)$ of $P$. The main result of this chapter is Theorem 16.1 of H. Cartan, which says that if the action admits a connection form, then there is a natural isomorphism between $\mathrm{H}_G^*(P)$ and the ordinary de Rham cohomology of the quotient space $M = P/G$. Restricting this isomorphism to the conjugacy invariant polynomials on $\mathfrak{g}$, one gets the Weil homomorphism of Theorem 13.1.

The algebra $A$ has a *double grading*: as a vector space it is equal to the direct sum of the spaces $A^{k,\,l}$ of the $\alpha \in A$, which are homogeneous polynomial mappings of degree $k$ from $\mathfrak{g}$, to the space $\Omega^l(P)$ of differential forms of degree $l$ on $P$. The *total degree* of $\alpha \in A^{k,\,l}$ is the number $m = 2k + l$; this is natural in view of the substitution of $X \in \mathfrak{g}$ by the curvature two-from $\Omega$ in Theorem 13.1. The space of equivariant forms of total degree $m$ will be denoted by

$$A^m := \bigoplus_{k,\,l \mid 2k+l=m} A^{k,\,l}. \tag{16.4}$$

The *equivariant exterior differentiation* $\mathrm{d}_{\mathfrak{g}}$ in the algebra $A$ is defined as the following combination of the exterior differentiation of differential forms and inner product with the infinitesimal action of $\mathfrak{g}$ in $P$:

$$(\mathrm{d}_{\mathfrak{g}}\,\alpha)(X) = \mathrm{d}\,(\alpha(X)) - \mathrm{i}(X_P)\,(\alpha(X)), \quad X \in \mathfrak{g},\ \alpha \in A. \tag{16.5}$$

We say that $\alpha \in A$ is *equivariantly closed* if $\mathrm{d}_{\mathfrak{g}}\,\alpha = 0$, and *equivariantly exact* if there exists a $\beta \in A$, such that $\alpha = \mathrm{d}_{\mathfrak{g}}\,\beta$.

If in (16.3) we replace $g$ by $\mathrm{e}^{-tX}$ and differentiate with respect to $t$ at $t = 0$ in the direction of $Y \in \mathfrak{g}$, then we get

$$0 = \mathcal{L}(X_P)\,(\alpha(X)) = \{\mathrm{d} \circ \mathrm{i}(X_P) + \mathrm{i}(X_P) \circ \mathrm{d}\}\,(\alpha(X)),$$

where the second identity is the homotopy formula for the Lie derivative of differential forms. Using this, one verifies that $\mathrm{d}_{\mathfrak{g}} \circ \mathrm{d}_{\mathfrak{g}} = 0$, which means that the equivariantly exact forms constitute a linear subspace of the space of equivariantly closed forms. So we can define the *equivariant cohomology* of the $G$-action in $P$ as

$$\mathrm{H}_G^*(P) := \ker \mathrm{d}_{\mathfrak{g}} \,/\, \mathrm{range}\, \mathrm{d}_{\mathfrak{g}} \,. \tag{16.6}$$

As usual, the equivariant cohomology class of $\alpha \in A$ will be denoted by

$$[\alpha] = \alpha + \mathrm{d}_{\mathfrak{g}}(A).$$

Clearly $\mathrm{d}_{\mathfrak{g}}$ maps $A^{k,\,l}$ to $A^{k,\,l+1} + A^{k+1,\,l-1}$, hence $A^m$ to $A^{m+1}$. Therefore, if $\mathrm{d}_{\mathfrak{g}}^m$ denotes the restriction of $\mathrm{d}_{\mathfrak{g}}$ to $A^m$, then

$$\mathrm{d}_{\mathfrak{g}}^m : A^m \to A^{m+1}. \tag{16.7}$$

So we can define the equivariant cohomology group of total degree $m$, of the $G$-action in $P$, as

$$\mathrm{H}_G^m(P) := \ker \mathrm{d}_{\mathfrak{g}}^m \,/\, \mathrm{range}\, \mathrm{d}_{\mathfrak{g}}^{m-1} \,. \tag{16.8}$$

Note that

$$\mathrm{H}_G^*(P) = \bigoplus_{m \in \mathbf{Z}_{\geq 0}} \mathrm{H}_G^m(P). \tag{16.9}$$

With the product of pointwise exterior multiplication of polynomial mappings $\mathfrak{g} \to \Omega^*(P)$, we have that

$$A^{k,\,l}\, A^{k',\,l'} \subset A^{k+k,\,l+l'},$$

hence

$$A^m\, A^{m'} \subset A^{m+m'}.$$

Since

$$\mathrm{d}_{\mathfrak{g}}(\alpha \wedge \beta) = (\mathrm{d}_{\mathfrak{g}}\, \alpha) \wedge \beta + (-1)^m \alpha \wedge \mathrm{d}_{\mathfrak{g}}\, \beta$$

if $\alpha \in A^m$, and $(-1)^m = (-1)^l$ if $m = 2k + l$, we get that $\alpha \wedge \beta$ is equivariantly closed, if both $\alpha$ and $\beta$ are equivariantly closed. And that $\alpha \wedge \beta$ is equivariantly exact if in addition either $\alpha$ or $\beta$ is equivariantly exact. It follows that $\mathrm{H}_G^*(M)$ has a uniquely defined product defined by

$$[\alpha] \cdot [\beta] = [\alpha \wedge \beta].$$

With this product, (16.9) is a graded algebra in the sense that $a \cdot b \in \mathrm{H}_G^{m+m'}(M)$ if $a \in \mathrm{H}_G^m(M)$ and $b \in \mathrm{H}_G^{m'}(M)$.

An important special case, and the only one needed for Theorem 13.1, occurs if the action of $G$ on $P$ is proper and free. This is equivalent to the condition that the orbit space $P/G$ is a smooth manifold $M$, and the projection

$$\pi : p \mapsto G \cdot p : P \to M$$

exhibits $P$ as a principal $G$-bundle over $M$. If $a$ is a differential form in $M$, then $\alpha = \pi^* a$ is a $G$-invariant differential form in $P$, which moreover is *horizontal* in the sense that

$$\mathrm{i}\,(X_P)\,\alpha = 0, \quad X \in \mathfrak{g}. \tag{16.10}$$

Moreover, the surjectivity of the $\mathrm{T}_p\,\pi$, $p \in P$ implies that the pullback operator $\pi^* : \Omega^*(M) \to \Omega^*(P)$ is injective. It follows that $\pi^*$ is an isomorphism from $\Omega^*(M)$ onto the space of $G$-invariant and horizontal differential forms in $P$.

The condition for $\alpha \in \Omega^*(P)$, of being $G$-invariant and horizontal, is well-defined also if it is not assumed that the action of $G$ is proper or free. Such $\alpha$ will be called *basic*, and the algebra of basic differential forms in $P$ will be denoted by $\Omega^*_{\mathrm{bas}}(P)$. The exterior differentiation leaves $\Omega^*_{\mathrm{bas}}(P)$ invariant, so we can form the *basic cohomology group*

$$\mathrm{H}^*_{\mathrm{bas}}(P) := \ker \mathrm{d} \cap \Omega^*_{\mathrm{bas}}(P) / \,\mathrm{d}\,(\Omega^*_{\mathrm{bas}}(P))\,. \tag{16.11}$$

Note that $\Omega^*_{\text{bas}}(P)$ is a subalgebra of $A$, and that d is equal to the restriction of $\mathrm{d}_{\mathfrak{g}}$ to $\Omega^*_{\text{bas}}(P)$, because $\mathrm{i}(X_P) = 0$ on elements of $A$ which are constant as polynomials on $\mathfrak{g}$. It follows that the mapping

$$\alpha + \mathrm{d}\left(\Omega^*_{\text{bas}}(P)\right) \mapsto \alpha + \mathrm{d}_{\mathfrak{g}}(A)$$

leads to a homomorphism

$$\imath_{\text{bas}} : \mathrm{H}^*_{\text{bas}}(P) \to \mathrm{H}^*_G(P). \tag{16.12}$$

If $P$ is a principal $G$-bundle, then the fact that the operators of exterior differentiation and pullback by $\pi$ commute implies that the isomorphism $\pi^* : \Omega^*(M) \to \Omega^*_{\text{bas}}(P)$ induces an isomorphism

$$\pi^* : \mathrm{H}^*(M) \xrightarrow{\sim} \mathrm{H}^*_{\text{bas}}(P) \tag{16.13}$$

from the de Rham cohomology of $M$ onto the basic cohomology of $P$.

## 16.2 Existence of a Connection Form

For a general action, a *connection form* is defined as a smooth $\mathfrak{g}$-valued one-form $\theta$ in $P$, which is $G$-equivariant in the sense that for each $g \in G$

$$\left(g_P^*\right)^{-1}\theta = (\operatorname{Ad} g)\ \theta, \tag{16.14}$$

and which reproduces $X \in \mathfrak{g}$ when applied to $X_P$:

$$\theta_p\left(X_{P,p}\right) = X, \quad p \in P. \tag{16.15}$$

Note that (16.15) implies that the action of $G$ in $P$ is *infinitesimally free* in the sense that if $X \in \mathfrak{g}$, $X \neq 0$, then $X_{P,p} \neq 0$ for each $p \in P$. In turn, this is equivalent to the condition that the action of $G$ in $P$ is *locally free*, in the sense that for each $p \in P$, there exists a neighborhood $U$ of $p$ in $P$ and a neighborhood $V$ of 1 in $G$ such that, if $p' \in U$, $g \in G$ and $g_P(p') = p'$, then $g = 1$.

Conversely, if the action of $G$ in $P$ is proper and locally (infinitesimally) free, then it has a connection form. The proof starts with the observation that the properness of the action implies that for every $p \in P$ there is a *slice through* $p$ for the $G$-action. This is a smooth submanifold $S$ of $P$ through $p$, such that

(i) $\mathrm{T}_p P$ is equal to the direct sum of $\mathrm{T}_p S$ and $\mathrm{T}_p(G \cdot p)$, the tangent space at $p$ of the orbit $G \cdot p$ of $p$.

(ii) $S$ is invariant under $G_p$, the stabilizer subsgroup $\{g \in G \mid g_P(p) = p\}$ of $p$ in $G$.

(iii) If $s \in S$, then $G_s \subset G_p$.

It follows that

$$U = G \cdot S = \{g_P(s) \mid g \in G,\, s \in S\}$$

is an open, $G$-invariant neighborhood of $p$ in $P$. And that the mapping $(g,\, p) \mapsto g_P(p)$ induces a diffeomorphism from $Q$ onto $U$ in which

$$Q := G \times_{G_p} S \tag{16.16}$$

is the orbit space of $G \times S$, for the proper and free action

$$h \mapsto \Big((g,\, s) \mapsto \big(g\, h^{-1},\, h_P(s)\big)\Big)$$

of $G_s$. The action of $G$ in $Q$, induced by the left multiplication of elements of $G$, acting on the first factor in $G \times S$, is transformed by the diffeomorphism $Q \to U$ into the $G$-action in $U$. The properness of the action also implies that the stabilizer group $G_p$ is compact. The condition that the action is locally free then makes it finite, so the projection

$$c : G \times S \to Q$$

is a finite covering. Obviously $G \times S$ has a unique connection form $\theta_{G\times S}$ for the left $G$-action, which is equal to zero on $0 \times \mathrm{T}\,S$. In turn, $\theta_{G\times S} = c^*\theta$ for a unique $\mathfrak{g}$-valued one-form $\theta$ in $Q$, and $\theta$ is a connection form in $Q$. Finally, the connection forms in the open $G$-invariant neighborhoods $U$ of $G$-orbits in $P$ are pieced together to a connection form in $P$ by means of a smooth $G$-invariant partition of unity (the construction of which also uses the local models $G \times_{G_s} S$ of the $U$'s).

In other words, if the action of $G$ in $P$ is proper, then the existence of a connection form is equivalent to the condition that the action is locally free. This condition is slightly weaker than the condition that $P$ is a principal $G$-bundle, that is, the action is proper and free. If the action is locally free but not free, then the orbit space $M = P/G$ no longer is a smooth manifold, but an orbifold, cf. Section 14.1. In this case the algebra of smooth differential forms in $M$ can be identified with the algebra $\Omega^*_{\mathrm{bas}}(P)$ of basic differential forms in $P$, introduced above for an arbitrary $G$-action in $P$. Proper actions which are locally free, but not free, occur sufficiently often to make it worthwhile to extend the theory to this case.

## 16.3 Henri Cartan's Theorem

Theorem 16.1 below states that, if the $G$-action in $P$ admits a connection form $\theta$, then the equivariant cohomology of $P$ is isomorphic to the basic cohomology of $P$. For the description of the basic form which is equivariantly cohomologous to a given equivariantly closed form, we will use a basis $e_i$ of $\mathfrak{g}$, and we write $X_i$ for the $i$-th coordinate of $X \in \mathfrak{g}$ with respect to this basis. Let $\Omega$ be the *curvature form*, the $\mathfrak{g}$-valued two-form in $P$ which is defined by (6.7) in terms of the connection form $\theta$. Then the coordinates $\Omega_i$ of $\Omega$ are scalar-valued two-forms in $P$, and it is to be expected that these can be expressed in terms of the one-forms $\theta_j$, the coordinates of $\theta$. Indeed, in terms of the basis $e_i$ of $\mathfrak{g}$, the Lie bracket of $\mathfrak{g}$ is given by

$$[e_i,\, e_j] = \sum_k c_{ij}^k\, e_k, \tag{16.17}$$

the coefficients $c_{ij}^k$ in which are called the *structure constants* of the Lie bracket, with respect to the basis $e_i$. Since $c_{ij}^k = -c_{ji}^k$ and

$$(\theta_i \wedge \theta_j)\,(v,\, w) = \theta_i(v)\,\theta_j(w) - \theta_i(w)\,\theta_j(v),$$

we get

$$\Omega_k = \mathrm{d}\,\theta_k - \tfrac{1}{2}\sum_{i,j} c_{ij}^k\,\theta_i \wedge \theta_j, \quad 1 \leq k \leq \dim \mathfrak{g}. \tag{16.18}$$

The equations (16.18) are called *Élie Cartan's structure equations*. The sum with the structure constants of $\mathfrak{g}$ in it is added in order to make the two-forms $\Omega_k$ horizontal, that is, $\mathrm{i}(X_P)\Omega_k = 0$ for each $X \in \mathfrak{g}$. In general, the $\Omega_k$ are not $G$-invariant, if the adjoint action of $G$ on $\mathfrak{g}$ is nontrivial, so the $\Omega_k$ need not be basic.

Any $\alpha(X) \in A$ can be written as

$$\alpha(X) = \sum_\mu \left(\prod_i X_i{}^{\mu_i}\right)\,\alpha_\mu,$$

in which the $\mu$ are multi-indices and the $\alpha_\mu$ are differential forms in $P$. The differential form

$$\alpha(\Omega) = \sum_\mu \left(\prod_i \Omega_i{}^{\mu_i}\right)\,\alpha_\mu$$

in $P$ is independent of the choice of the basis in $\mathfrak{g}$. Note that the order of the factors $\Omega_i$ does not matter, because two-forms commute.

Finally, every $\alpha \in A$ has a unique decomposition

$$\alpha = \alpha_{\text{hor}} + \sum_{j=1}^{\dim \mathfrak{g}} \sum_{i_1 < \ldots < i_j} \theta_{i_1} \wedge \ldots \theta_{i_j} \wedge \beta_{i_1 \ldots i_j}, \tag{16.19}$$

in which $\alpha_{\text{hor}}$ and $\beta_{i_1 \ldots i_j}$ are horizontal elements of $A$. We will call $\alpha_{\text{hor}}$ the *horizontal part* of $\alpha$.

**Theorem 16.1** *If there exists a connection form $\theta$ for the $G$-action in $P$, then the homomorphism (16.12), from the basic cohomology of $P$ to the equivariant cohomology $\mathrm{H}^*_G(P)$, is an isomorphism. If $\pi : P \to M$ is a principal fiber bundle, then the composition of $\pi^*$ with (16.12) is an isomorphism, from the de Rham cohomology $\mathrm{H}^*(M)$ of $M$ onto the equivariant cohomology $\mathrm{H}^*_G(P)$ of $P$.*

*Secondly, let $\alpha(X)$ be an equivariantly closed form. Then*

$$(\alpha_{\mathrm{hor}})\,(\Omega) = (\alpha(\Omega))_{\mathrm{hor}}\,, \tag{16.20}$$

*and this form is basic, closed, and equivariantly cohomologous to $\alpha(X)$.*

**Proof** We will work out the proof which is indicated by the sentence

"On raisonne alors comme dans la démonstration du théorème 1, ..."

in [16, p. 63]. If $\alpha$ is a polynomial mapping from $\mathfrak{g}$ to $\Omega^*(P)$, then the polynomial mapping $\partial_i\alpha$ from $\mathfrak{g}$ to $\Omega^*(P)$ is defined by

$$(\partial_i\alpha)(X) := \tfrac{d}{dt}\alpha(X + t\,e_i)|_{t=0}.$$

We also write

$$v_i := e_{i,\,P} = \tfrac{d}{dt}\,(\exp t\,e_i)_P\,|_{t=0}$$

for the vector field in $P$ which is equal to the infinitesimal action in $P$ of the element $e_i$ in the Lie algebra of $G$.

Let $\alpha \in A$, that is, $\alpha$ is a $G$-equivariant polynomial mapping from $\mathfrak{g}$ to $\Omega^*(P)$. We define the mapping $k\,\alpha : \mathfrak{g} \to \Omega^*(P)$ by

$$(k\,\alpha)(X) = -\sum_i \theta_i \wedge \partial_i\alpha(X). \tag{16.21}$$

One can verify directly that $k\,\alpha$ is independent of the choice of the basis in $\mathfrak{g}$ and, more importantly, belongs to $A$ again. Furthermore, the linear mapping $k : A \to A$ maps $A^{k,l}$ to $A^{k-1,l+1}$: it increases the form degree by one and decreases the the degree as a polynomial on $\mathfrak{g}$ by one. It therefore maps $A^m$ to $A^{m-1}$. Also note that $k \circ k = 0$, because $\partial_i \circ \partial_j$ is symmetric in $i$, $j$ and $\theta_i \wedge \theta_j$ is antisymmetric in $i$, $j$.

Combining $k$ and $\mathrm{d}_{\mathfrak{g}}$, we get the "homotopy operator"

$$h := \mathrm{d}_{\mathfrak{g}} \circ k + k \circ \mathrm{d}_{\mathfrak{g}} : A^m \to A^m. \tag{16.22}$$

Note that $\mathrm{d}_{\mathfrak{g}} \circ \mathrm{d}_{\mathfrak{g}} = 0$ and $k \circ k = 0$ imply that

$$h \circ \mathrm{d}_{\mathfrak{g}} = \mathrm{d}_{\mathfrak{g}} \circ k \circ \mathrm{d}_{\mathfrak{g}} = \mathrm{d}_{\mathfrak{g}} \circ h,$$

and

$$h \circ k = k \circ \mathrm{d}_{\mathfrak{g}} \circ k = k \circ h,$$

so $h$ commutes with $\mathrm{d}_{\mathfrak{g}}$ and with $k$.

A direct calculation shows that

$$h = -S - C + E + V, \tag{16.23}$$

in which the operators $S$, $C$, $E$ and $V$ in $A^m$ are defined by

$$(S\,\alpha)(X) = \sum_i \Omega_i \wedge \partial_i \alpha(X), \tag{16.24}$$

$$(C\,\alpha)(X) = \tfrac{1}{2} \sum_{i,j,k} c_{ij}^k \theta_i \wedge \theta_j \wedge \partial_k \alpha(X), \tag{16.25}$$

$$(E\,\alpha)(X) = \sum_i X_i\, \partial_i \alpha(X), \tag{16.26}$$

$$(V\,\alpha)(X) = \sum_i \theta_i \wedge \mathrm{i}\,(v_i)\; \alpha(X). \tag{16.27}$$

In order to investigate the invertibility properties of $h$, we introduce the filtration in $A$ by the degree as a polynomial on $\mathfrak{g}$:

$$A_{\leq k} := \bigoplus_{r,\,l \mid r \leq k} A^{r,\,l}. \tag{16.28}$$

Then $S$ and $C$, in (16.24) and (16.25) respectively, map $A_{\leq k}$ into $A_{\leq (k-1)}$, so are equal to zero on $A_{\leq k}/A_{\leq (k-1)}$.

We say that $\alpha \in A$ has *verticality equal to* $j$, if in the decomposition (16.19) only terms occur with $j$ factors $\theta_i$. We write $A^{j,\,k,\,l}$ for the space of $\alpha(X) \in A$, which have verticality $j$, degree $k$ as a polynomial on $\mathfrak{g}$, and degree $l$ as a differential form in $P$. Note that the previously introduced space $A^{k,\,l}$ is equal to the direct sum over $j$ of the spaces $A^{j,\,k,\,l}$. Also, the sum over $l$ of the spaces $A^{0,\,0,\,l}$ is equal to the algebra $\Omega^*_{\mathrm{bas}}(P)$ of basic differential forms in $P$.

The Euler operator $E$ in (16.26) acts on $A^{j,\,k,\,l}$ as multiplication by $k$. Using that

$$\mathrm{i}(v_p) \circ \mathrm{e}(\theta_q) = \langle v_p,\, \theta_q \rangle - \mathrm{e}(\theta_q) \circ \mathrm{i}(v_p)$$

and $\langle v_p,\, \theta_q \rangle = \delta_{pq}$, one gets that the operator $V$ in (16.27) acts on $A^{j,\,k,\,l}$ as multiplication by $j$. The conclusion is that $h = -S - C + E + V$ acts on $A_{\leq k}/A_{\leq (k-1)}$ as a diagonalizable operator, with eigenvalues equal to $k + j$, $0 \leq j \leq \dim \mathfrak{g}$. In particular, for every $k \geq 1$, $h$ is invertible on $A_{\leq k}/A_{\leq (k-1)}$.

This in turn implies that

$$h : A/A_{\leq 0} \to A/A_{\leq 0}$$

is invertible. Write $N = \ker \mathrm{d}_{\mathfrak{g}}$. Since $h$ commutes with $\mathrm{d}_{\mathfrak{g}}$, it follows that $h$ induces an invertible transformation $h_0$ in $N/N \cap A_{\leq 0}$. Also note that $h = \mathrm{d}_{\mathfrak{g}} \circ k$ on $N$. The conclusion is that for every $\alpha \in N$, there exists a $\nu \in N$ and a $\lambda \in N \cap A_{\leq 0}$, such that

$$\alpha = h\,\nu + \lambda = \mathrm{d}_{\mathfrak{g}}\; k\,\nu + \lambda, \tag{16.29}$$

or the equivariant cohomology class of $\alpha$ is equal to the one of $\lambda \in A_{\leq 0}$. For $\lambda \in A_{\leq 0}$, the condition $\mathrm{d}_{\mathfrak{g}}\,\lambda = 0$ means that $\mathrm{d}\lambda = 0$ and, for each $X \in \mathfrak{g}$, $\mathrm{i}\,(X_P)\;\lambda = 0$. That is, $\lambda$ is a closed element of $\Omega^*_{\mathrm{bas}}(P)$. The conclusion is, that the surjectivity of $h_0$ implies that the homomorphism (16.12) from $\mathrm{H}^*_{\mathrm{bas}}(P)$ to $\mathrm{H}^*_G(P)$ is surjective.

In order to prove the injectivity of $i_{\mathrm{bas}}$, let $\alpha$ be a basic differential form, such that

$$i_{\mathrm{bas}}\left([\alpha]_{\mathrm{bas}}\right) = 0,$$

if $[\alpha]_{\mathrm{bas}}$ denotes the cohomology class in $\mathrm{H}^*_{\mathrm{bas}}(P)$ of $\alpha$. This means, that there exists a $\beta(X) \in A$, such that

$$\alpha = (\mathrm{d}_{\mathfrak{g}}\, \beta)\,(X) = \mathrm{d}\,(\beta(X)) - \mathrm{i}(X_P)\,(\beta(X))$$

Substituting $X = 0$, we get that

$$\alpha = \mathrm{d}\,\beta(0) = \mathrm{d}_{\mathfrak{g}}\,\beta(0),$$

in which $\beta(0) \in A_{\leq 0}$. Now we observe that $S$ and $C$ vanish on $A_{\leq 0}$, and that $h$ acts on $A^{j,\,0,\,l}$ as multiplication by $j$. It follows that the operator

$$P = \prod_{j=1}^{\dim \mathfrak{g}} (j - h)$$

is equal to zero on $A^{j,\,0,\,l}$ if $j \geq 1$, and equal to multiplication by $(\dim \mathfrak{g})!$ on $A^{0,\,0,\,l}$. In other words,

$$H = \frac{1}{(\dim \mathfrak{g})!}\, P$$

is equal to the linear projection from $A_{\leq 0}$ onto the subspace of the basic forms, along the spaces $A^{j,\,0,\,l}$ with $j > 0$. Since $h$ commutes with $\mathrm{d}_{\mathfrak{g}}$, also $H$ commutes with $\mathrm{d}_{\mathfrak{g}}$, and we now get, because $\alpha$ is basic:

$$\alpha = H\,\alpha = \mathrm{d}_{\mathfrak{g}}\,(H\,\beta(0)) = \mathrm{d}\,(H\,\beta(0))$$

This shows that $\alpha$ is equal to the exterior derivative of a basic form, which means that its cohomology class in $\mathrm{H}^*_{\mathrm{bas}}(P)$ is equal to zero. This completes the proof that $i_{\mathrm{bas}}$ is injective.

In order to determine the basic form $\lambda$ in (16.29) in terms of $\alpha$, we observe that the fact that the $\Omega_i$ are horizontal, implies that S maps $A^{j,\,k,\,l}$ to $A^{j,\,k-1,\,l+2}$. On the other hand, C maps $A^{j,\,k,\,l}$ to $A^{j+2,\,k-1,\,l+2}$. Recall also that $E$ and $V$ act in $A^{j,\,k,\,l}$ as multiplication by $k$ and $j$, respectively.

Write $\alpha^{j,\,k,\,l}$ for the $A^{j,\,k,\,l}$-component of any $\alpha \in A$, that is,

$$\alpha = \sum_{j,\,k,\,l} \alpha^{j,\,k,\,l}, \quad \alpha^{j,\,k,\,l} \in A^{j,\,k,\,l}.$$

The equation (16.29), $\alpha = h\,\nu + \lambda$, then reads, written out in components:

$$\alpha^{j,\,k,\,l} = -S\,\nu^{j,\,k+1,\,l-2} - C\,\nu^{j-2,\,k+1,\,l-2} + (k+j)\,\nu^{j,\,k,\,l} \tag{16.30}$$

if $(j,\,k) \neq (0,\,0)$, and

$$\alpha^{0,\,0,\,l} = -S\,\nu^{0,\,1,\,l-2} + \lambda^{0,\,0,\,l}. \tag{16.31}$$

The equation (16.30) for $j = 0$ can be written as

$$\nu^{0,\,k,\,l} = \tfrac{1}{k}\,S\,\nu^{0,\,k+1,\,l} + \tfrac{1}{k}\,\alpha^{0,\,k,\,l}, \quad k \geq 1.$$

Since $\nu^{j,\,k,\,l} = 0$ if $k$ is greater than the polynomial degree of $\nu$, this leads to a unique determination of $\nu^{0,\,1,\,l}$ in terms of $\alpha$:

$$\nu^{0,\,1,\,l} = \sum_{k \geq 1} \tfrac{1}{k!}\,S^{k-1}\,\alpha^{0,\,k,\,l-2(k-1)}.$$

If we now combine this with (16.31), we get

$$\lambda^{0,\,0,\,l} = \sum_{k \geq 0} \tfrac{1}{k!}\,S^{k}\,\alpha^{0,\,k,\,l-2k}.$$

But this just means that

$$\lambda = (\alpha_{\mathrm{hor}})\,(\Omega),$$

because

$$\beta(\Omega) = \sum_{k \geq 0} \tfrac{1}{k!}\,S^{k}\,\beta, \quad \beta \in A. \tag{16.32}$$

Since $S$ commutes with $V$, and $\alpha \mapsto \alpha_{\mathrm{hor}}$ is equal to the linear projection onto the kernel of $V$ along the sum of the eigenspaces of $V$ for the nonzero eigenvalues, we get that

$$S\,(\alpha_{\mathrm{hor}}) = (S\,\alpha)_{\mathrm{hor}}\,.$$

Combining this with (16.32), we get (16.20). □

Note that (16.30) also leads to an explicit determination, in terms of $\omega$, of all $\nu^{j,k,l}$ with $(j, k) \neq (0, 0)$. This leaves the freedom in $\nu$ of adding a basic form $\tilde{\nu}$. Since $k$ acts as zero on the space of basic forms, we get a full explicit determination of the equivariant form $k\,\nu$ in (16.29).

## 16.4 Proof of Weil's Theorem

For the proof of Theorem 13.1, we apply Theorem 16.1 to $\alpha(X) = f(X)$, for some $f \in \mathbf{C}[\mathfrak{g}]^{\mathrm{Ad}\,G}$. That is, $\alpha$ is an equivariantly closed form which is a constant, when regarded as a differential form in $P$. We get that $f(\Omega)$ is a closed, basic form, and equivariantly cohomologous to $f(X)$. Since also $f(\tilde{\Omega})$ is equivariantly cohomologous to $f(X)$, we get that $f(\tilde{\Omega}) - f(\Omega)$ is equivariantly cohomologous to zero. The injectivity of $i_{\mathrm{bas}}$ implies now that there exists a basic form $\beta$ in $P$ such that $f(\tilde{\Omega}) - f(\Omega)$ is equal to the exterior derivative of $\beta$.

## 16.5 General Actions

We conclude this chapter with the remark that in general equivariant cohomology has "very big" contributions from $G$-invariant neighborhoods of points $p \in P$, where the stabilizer subgroup $G_p$ is not discrete. Suppose that the $G$-action in $P$ is proper. Then the compactness of $G_p$ implies that, by restricting $S$ to a suitable neighborhood of $p$, there are local coordinates around $p$ in $S$ which the action of $G_p$ is linear. The radial contraction in $S$ in the local model (16.16) then commutes with the $G$-action, so we have a $G$-equivariant retraction of $U$ to the orbit $G \cdot p \simeq G/G_p$ through $p$. This leads to the conclusion that

$$\mathrm{H}_G^*(U) \simeq \mathrm{H}_G^*(G/G_p) \simeq \mathbf{C}[\mathfrak{g}_p]^{\mathrm{Ad}\,G_p}, \tag{16.33}$$

the ring of the Ad $G_p$-invariant polynomials on $\mathfrak{g}_p$. As soon as $\dim \mathfrak{g}_p > 0$, this ring has nonzero contributions in arbitrarily high degree. Using Mayer-Vietoris sequences as in Bott and Tu [14, Ch. II], one can conclude that, as soon as the action of $G$ in $P$ is not locally free, then $\mathrm{H}_G^m(P) \neq 0$ for arbitrarily large $m$. This is in strong contrast with ordinary de Rham cohomology. Related to this is the infinite-dimensionality of the classifying spaces which occurs in the topological definition of equivariant cohomology, cf. Atiyah and Bott [6]. Also related to this is the localization formula at the fixed points of a torus action, of Berline-Vergne [10] and Atiyah-Bott [6], for integrals of equivariant forms. See also Berline, getzler and Vergne [9, Ch. 7].

# Bibliography

[1] R. Abraham and J.E. Marsden: *Foundations of Mechanics*. The Benjamin/Cummings Publ. Co., Reading, Mass., 1987.

[2] M.F. Atiyah: *Elliptic Operators and Compact Groups*. Lecture Notes in Math. **401**, Springer-Verlag, Berlin, 1974.

[3] M.F. Atiyah: Convexity and commuting Hamiltonians. *Bull. London Math. Soc.* **14** (1982), 1-15.

[4] M.F. Atiyah, R. Bott, and A. Shapiro: Clifford modules. *Topology* **3** Suppl. 1 (1964), 3-38.

[5] M.F. Atiyah and R. Bott: A Lefschetz fixed point formula for elliptic complexes: II. Applications. *Ann. of Math.* **88** (1968), 451-491.

[6] M.F. Atiyah and R. Bott: The moment map and equivariant cohomology. *Topology* **23** (1984), 1-28.

[7] M.F. Atiyah and I.M. Singer: The index of elliptic operators III. *Ann. of Math.* **87** (1968), 546-604.

[8] F.A. Berezin: *The Method of Second Quantization*. Pure and Applied Physics 24, Academic Press, New York, London, 1966.

[9] N. Berline, E. Getzler, and M. Vergne: *Heat Kernels and Dirac Operators*. Springer-Verlag, Berlin, 1992.

[10] N. Berline and M. Vergne: Classes caractéristiques équivariantes. Formules de localisation en cohomologie équivariante. *C. R. Acad. Sci. Paris* **295** (1982), 539-541.

[11] N. Berline and M. Vergne: A computation of the equivariant index of the Dirac operator. *Bull. Soc. Math. France* **113** (1985), 305-345.

[12] S. Bochner: Compact groups of differentiable transformations. *Ann. of Math.* **46** (1945), 372-381.

[13] R. Bott: A residue formula for holomorphic vector fields. *J. Diff. Geom.* **1** (1967), 311-330.

[14] R. Bott and L.W. Tu: *Differential Forms in Algebraic Topology.* Springer-Verlag, New York, Heidelberg, Berlin, 1982.

[15] R. Brauer and H. Weyl: Spinors in $n$ dimensions. *Amer. J. Math.* **57** (1935), 425-449.

[16] H. Cartan: Notions d'algèbre différentielle; applications aux variétés où opère un groupe de Lie. Also: La transgression dans un groupe de Lie et dans un fibré principal. pp. 15-27 and 57-71 in: *Colloque de Topologie*. C.B.R.M., Bruxelles, 1950.

[17] W.K. Clifford: Applications of Grassmann's extensive algebra. *Amer. J. Math.* **1** (1878), 350-358.

[18] P.A.M. Dirac: The quantum theory of the electron. *Proceedings of the Royal Society (A)* **117** (1927) 610-624. Part II in **118** (1928), 351-361.

[19] J.J. Duistermaat: On operators of trace class in $L^2(X, \mu)$. Geometry and Analysis. Papers dedicated to the memory of V.K. Patodi. Springer-Verlag 1981, pp. 29-32; *Proc. Indian Acad. Sci. Math. Sci.* **90** (1981), 29-32.

[20] J.J. Duistermaat: The momentum map. pp. 347-392 in ed. J. Szenthe and L. Tamássy: *Topics in Differential Geometry*, Vol. I. North-Holland Publ. House, Amsterdam, 1988.

[21] J.J. Duistermaat: Equivariant cohomology and stationary phase. *Contemporay Mathematics* **179** (1994), 45-62. Amer. Math. Soc., Providence, Rhode Island.

[22] J.J. Duistermaat: *Fourier Integral Operators*. Birkhäuser, Boston, 1995.

[23] J.J. Duistermaat and G.J. Heckman: On the variation in the cohomology of the symplectic form of the reduced phase space. *Invent. math.* **69** (1982), 259-268. Addendum in loc. cit. **72** (1983), 153-158.

[24] J.J. Duistermaat, J.A.C. Kolk and V.S. Varadarajan: Functions, flows and oscillatory integrals on flag manifolds and conjugacy classes in real semisimple Lie groups. *Compositio Mathematica* **49** (1983), 309-398.

[25] D. van Dantzig and B.L. van der Waerden: Über metrisch homogene Räume. *Abh. Math. Sem. Univ. Hamburg* **6** (1928), 367-376.

[26] J. Eichhorn and T. Friedrich: Seiberg-Witten theory. To appear in: *Banach Center Publications*, Warsaw.

[27] E. Getzler: Pseudodifferential operators on supermanifolds and the Atiyah-Singer index theorem. *Comm. Math. Phys.* **92** (1983), 163-178.

[28] P.B. Gilkey: Curvature and the eigenvalues of the Dolbeault complex for Kaehler manifolds. *Adv. in Math.* **11** (1973), 311-325.

[29] P.B. Gilkey: Spectral geometry and the Kaehler condition for complex manifolds. *Invent. math.* **26** (1974), 231-258. Correction in ibid. **29** (1975), 81-82.

[30] P.B. Gilkey: The spectral geometry of a Riemannian manifold. *J. Diff. Geom.* **10** (1975), 601-618.

[31] P.B. Gilkey: Curvature and the eigenvalues for the Dolbeault complex for Hermitian manifolds. *Adv. in Math.* **21** (1976), 61-77.

[32] P.B. Gilkey: *Invariance Theory, the Heat Equation, and the Atiyah-Singer Index Theorem.* Publish or Perish, Inc., Wilmington, Delaware (U.S.A.), 1984.

[33] P. Griffiths and J. Harris: *Principles of Algebraic Geometry.* John Wiley & Sons, New York, 1978.

[34] V. Guillemin: *Moment Maps and Combinatorial Invariants of Hamiltonian $T^n$-spaces.* Birkhäuser, Boston, 1994.

[35] V. Guillemin and S. Sternberg: Convexity properties of the moment mapping. *Invent. math.* **67** (1982), 491-513.

[36] V. Guillemin and S. Sternberg: Geometric quantization and multiplicities of group representations. *Invent. math.* **67** (1982), 515-538.

[37] J. Hadamard: *Lectures on Cauchy's Problem in Linear Partial Differential Equations.* Yale University Press, 1923. Reprinted in: Dover Publications, New York, 1952.

[38] S. Helgason: *Differential Geometry, Lie Groups and Symmetric Spaces.* Academic Press, Orlando, 1978.

[39] F. Hirzebruch: Arithmetic genera and the theorem of Riemann-Roch for algebraic varieties. *Proc. Nat. Acad. Sci. U.S.A.* **40** (1954), 110-114.

[40] F. Hirzebruch und H. Hopf: Felder von Flächenelementen in 4-dimensionalen Mannigfaltigkeiten. *Math. Annalen* **136** (1958), 156-172.

[41] L. Hörmander: *An Introduction to Complex Analysis in Several Variables*. North-Holland Publ. Co., Amsterdam, 1973.

[42] L. Hörmander: *The Analysis of Linear Partial Differential Operators III*. Springer-Verlag, Berlin, 1985.

[43] M.W. Hirsch, C.C. Pugh, M. Shub: *Invariant Manifolds*. Lecture Notes in Mathematics **583**. Springer-Verlag, Berlin, 1977.

[44] T. Kawasaki: The signature theorem for V-manifolds. *Topology* **17** (1978), 75-83.

[45] T. Kawasaki: The Riemann-Roch theorem for complex V-manifolds. *Osaka J. Math.* **16** (1979), 151-159.

[46] T. Kawasaki: The index of elliptic operators over V-manifolds. *Nagoya Math. J.* **84** (1981), 135-157.

[47] A.A. Kirillov: Unitary representations of nilpotent groups. *Uspekhi Mat. Nauk* **17, 4** (1962), 57-110. Translated in *Russ. Math. Surveys* **17, 4** (1962), 53-104.

[48] F.C. Kirwan: Convexity properties of the moment mapping III. *Invent. math.* **77** (1984), 547-552.

[49] K. Kodaira: On a differential-geometric method in the theory of analytic stacks. *Proc. Nat. Acad. Sci. U. S. A.* **39** (1953), 1268-1273.

[50] B. Kostant: Quantization and unitary representations. pp. 87-208 in: ed. C.T. Taam: *Lectures in Modern Analysis and Applications III*. Lecture Notes in Mathematics 170, Springer-Verlag, Berlin, 1970.

[51] H.B. Lawson and M.-L. Michelsohn: *Spin Geometry*. Princeton University Press, Princeton, New Jersey, 1989.

[52] A. Lichnerowicz: Spineurs harmoniques. *C. R. Acad. Sci. Paris*, Sér. A **257** (1963), 7-9.

[53] A. Lichnerowicz: Les variétés de Poisson et leurs algèbres de Lie associées. *J. Diff. Geom.* **12** (1977), 253-300.

[54] S. Lie: *Theorie der Transformationsgruppen II.* Teubner Verlag, Leipzig u. Berlin, 1890, reprinted in 1930.

[55] B. Malgrange: Sur l'integrabilité des structures presques-complexes. pp. 289-296 in: *Symposia Mathematica* II, INDAM, Rome, 1968.

[56] J. Marsden and A. Weinstein: Reduction of symplectic manifolds with symmetry. *Reports on Math. Phys.* **5** (1974), 121-130.

[57] H. McKean and I.M. Singer: Curvature and the eigenvalues of the Laplacian. *J. Diff. Geom.* **1** (1967), 43-69.

[58] E. Meinrenken: Symplectic surgery and the spin$^c$-Dirac operator. To appear in *Advances in Mathematics.*

[59] J.W. Milnor and J.D. Stasheff: *Characteristic Classes.* Annals of Math. Studies 76, Princeton University Press, Princeton, N.J., 1974.

[60] S. Minakshisundaram and Å. Pleijel: Some properties of the eigenfunctions of the Laplace operator on Riemannian manifolds. *Canad. J. Math.* **1** (1949), 242-256.

[61] J. W. Morgan: *The Seiberg-Witten Equations and Applications to the Topology of Smooth Four-Manifolds.* Notes of Lectures at Columbia University and Princeton University, Columbia University, 1995.

[62] A. Newlander and L. Nirenberg: Complex analytic coordinates in almost complex manifolds. *Ann. of Math.* **65** (1957), 391-404.

[63] V.K. Patodi: Curvature and the eigenforms of the Laplace operator. *J. Diff. Geom.* **5** (1971), 233-249.

[64] V.K. Patodi: An analytic proof of Riemann-Roch-Hirzebruch theorem for Kähler manifolds. *J. Diff. Geom.* **5** (1971), 251-283.

[65] V.K. Patodi: Holomorphic Lefschetz fixed point formula. *Bull. Amer. Math. Soc.* **79** (1973), 825-828.

[66] I. Satake: On a generalization of the notion of manifold. *Proc. Nat. Acad. Sci.* **42** (1956), 359-363.

[67] I. Satake: The Gauss-Bonnet for V-manifolds. *J. Math. Soc. Japan* **9** (1957), 464-492.

[68] J.-P Serre: Représentations linéaires et espaces homogènes kählериennes des groupes de Lie compacts. *Séminaire Bourbaki* 1953/54. Exposé 100. Secrétariat Mathématique, Paris.

[69] J.-P. Serre: Géometrique algébrique et géométrie analytique. *Ann. de l'Inst. Fourier* **6** (1955/56), 1-42.

[70] R. Sjamaar: Convexity properties of the moment mapping re-examined. To appear in *Advances in Mathematics*.

[71] J. Tits: Sur certaines classes d'espaces homogènes de groupes de Lie. *Acad. Roy. Belg. Cl. Sci. Mem. Coll.* **29** (1955), no. 3.

[72] S.R.S. Varadhan: On the behaviour of the fundamental solution of the heat equation with variable coefficients. *Comm. Pure Appl. Math.* **20** (1967), 431-455.

[73] M. Vergne: Equivariant index formula for orbifolds. Preprint No. 94-21, Laboratoire de Mathématiques de l'École Normale Supérieure, 45, rue d'Ulm, 75230 Paris Cedex 05, October 1994.

[74] A. Weil: (1949) Géometrie différentielle des espaces fibrés. *Œuvres Scientifiques (Collected Papers)*, vol. I, pp. 422-436. Springer-Verlag, New York, Heidelberg, Berlin, 1980.

[75] R. O. Wells, Jr.: *Differential Analysis on Complex Manifolds*. Prentice-Hall, Inc., Englewood Cliffs, N. J., 1973.

[76] Wu Wen-Tsun: Sur les classes caractéristiques des structures fibrées sphériques. pp. 5-89 in: Wu Wen-Tsun et G. Reeb: *Sur les Espaces Fibrés et les Variétés Feuilletées*. Hermann, Paris, 1952.

# Index

GPSR Compliance

*The European Union's (EU) General Product Safety Regulation (GPSR) is a set of rules that requires consumer products to be safe and our obligations to ensure this.*

*If you have any concerns about our products, you can contact us on ProductSafety@springernature.com*

In case Publisher is established outside the EU, the EU authorized representative is:

Springer Nature Customer Service Center GmbH
Europaplatz 3
69115 Heidelberg, Germany

**Batch number: 10577876**

Printed by Printforce, the Netherlands